Urban Geopolitics

In the last decade a new wave of urban research has emerged, putting comparative perspectives back on the urban studies agenda. However, this research is frequently based on similar case studies on a few selected cities in America and Europe and all too often focus on the abstract city level with marginal attention given to particular local contexts.

Moving away from loosely defined urban theories and contexts, this book argues it is time to start learning from and compare across different 'contested cities'. It questions the long-standing Euro-centric academic knowledge production that is prevalent in urban studies and planning research. This book brings together a diverse range of international case studies from Latin America, South and South East Asia, Eastern Europe, Africa and the Middle East to offer an in-depth understanding of the worldwide contested nature of cities in a wide range of local contexts. It suggests an urban ontology that moves beyond the urban 'West' and 'North' as well as adding a comparative-relational understanding of the contested nature that 'Southern' cities are developing.

This timely contribution is essential reading for those working in the fields of human geography, urban studies, planning, politics, area studies and sociology.

Jonathan Rokem, PhD, is Marie Skłodowska-Curie Research Fellow at the Bartlett School of Architecture, University College London (UCL), UK. His research interests and publications focus on spatial and social critical analysis of cities and regions.

Camillo Boano, PhD, is Senior Lecturer at the Bartlett Development Planning Unit and Director of the MSc in Building and Urban Design in Development, UCL, UK. He is author of *The Ethics of a Potential Urbanism: Critical Encounters Between Giorgio Agamben and Architecture* (2017).

Routledge Studies in Urbanism and the City

This series offers a forum for original and innovative research that engages with key debates and concepts in the field. Titles within the series range from empirical investigations to theoretical engagements, offering international perspectives and multidisciplinary dialogues across the social sciences and humanities, from urban studies, planning, geography, geohumanities, sociology, politics, the arts, cultural studies, philosophy and literature.

For a full list of titles in this series, please visit www.routledge.com/series/RSUC

Urban Geopolitics

Rethinking Planning in Contested Cities

Edited by Jonathan Rokem
and Camillo Boano

LONDON AND NEW YORK

First published 2018
by Routledge

2 Park Square, Milton Park, Abingdon, Oxfordshire OX14 4RN
52 Vanderbilt Avenue, New York, NY 10017

Routledge is an imprint of the Taylor & Francis Group, an informa business

First issued in paperback 2020

British Library Cataloguing-in-Publication Data
A catalogue record for this book is available from the British Library

Library of Congress Cataloging-in-Publication Data
A catalog record for this book has been requested

ISBN: 978-1-138-96266-8 (hbk)
ISBN: 978-0-367-66771-9 (pbk)

Typeset in Times New Roman
by Cenveo Publisher Services

Contents

Illustrations

Figures

Tables

Contributors

Apurba Kumar Podder has recently completed his PhD from the University of Cambridge. He examines how illegality as a condition informs and shapes the internal dynamics of informal growth in developing cities.

Camila Cociña is a Teaching Fellow and PhD candidate at the Bartlett Development Planning Unit, UCL. Her current research focuses on housing policies and urban inequalities in the Chilean context.

Camillo Boano, PhD, is Senior Lecturer at the Bartlett Development Planning Unit, UCL, and Director of the MSc in Building and Urban Design in Development. He is author of *The Ethics of a Potential Urbanism: Critical Encounters Between Giorgio Agamben and Architecture* (2017).

Catalina Ortiz, PhD, is Lecturer at the Bartlett Development Planning Unit, UCL. She is researching critical spatial practices intersecting urban design, land management, large-scale projects, strategic spatial planning and urban policy mobility in the Global South.

Ernesto López-Morales is Associate Professor in the University of Chile and Associate Researcher at the Centre for Social Conflict and Cohesion Studies (COES), where he focuses on land economic, gentrification, neoliberal urbanism and housing in Chile and Latin American cities.

Gruia Bădescu is a Lecturer in Human Geography at Christ Church, University of Oxford. His research interests include post-war reconstruction and coming to terms with the past.

James D. Sidaway is based at the National University of Singapore, where he is Professor of political geography. His research currently focuses on security and insecurity and the history and philosophy of geography.

Jonathan Rokem, PhD, is Marie Skłodowska-Curie Research Fellow at the Bartlett School of Architecture, UCL. His research interests and publications focus on spatial and social critical analysis of cities and regions.

Kayvan Karimi is a Senior Lecturer within UCL's Space Syntax Laboratory and Director of Space Syntax Limited, a UCL knowledge transfer spin-off. His main areas of research include evidence-based design and planning, organic and informal urbanism and large-scale urban systems.

Laura Vaughan is Professor of Urban Form and Society and Director of UCL's Space Syntax Laboratory. Laura's research addresses the inherent complexity of the urban environment both theoretically and methodologically. Her most recent book, *Suburban Urbanities*, was published by UCL Press in 2015.

Liza Rose Cirolia is a Researcher at the African Centre for Cities, University of Cape Town. Her research focuses on housing, infrastructure and land in African cities. Her current PhD research is focused on public finance in African cities.

Michael Safier is Professor Emeritus at the Development Planning Unit, UCL. He dedicated his life, research and teaching activities, and professional practice to urban planning development, specifically around the conceptualization of cosmopolitan planning.

Moriel Ram, PhD, is a Research Fellow at the Institute of Advanced Studies at UCL. His research interests are militarized geographies and urban geopolitics of faith in the Middle East and the Mediterranean.

Nimrod Luz is Associate Professor at the Sociology and Anthropology Department, Western Galilee College. His research interests include geography of religion with particular interest in minorities' sacred sites, changing cultural landscape of the Middle East and religiosity in the mixed town of Acre.

Nurit Stadler is Associate Professor at the Sociology and Anthropology Department, the Hebrew University of Jerusalem. Her research interests include Israel's Ultraorthodox community, fundamentalism, Greek Orthodox and Catholic rituals in Jerusalem.

Pawda F. Tjoa is a researcher at Publica, a London-based public realm consultancy. She recently completed her PhD at the University of Cambridge, during which she explored the connection between public space and political ideology in post-independence Jakarta.

Peter D. A. Wood is a Postdoctoral Fellow in Demography at the Federal University of Minas Gerais in Belo Horizonte, Brazil. His work focuses on migration, violence and political participation in Global South development projects.

Sadaf Sultan Khan has recently completed her PhD at the Space Syntax Laboratory, UCL. Her work focuses on the appropriation, adaptation and contestation of urban space by migrant communities.

Sara Fregonese is based at the University of Birmingham, where she is Research Fellow in urban resilience. Her research currently focuses on the urban geopolitics of civil war and terrorism and on historical-political geographies of sectarianism in the Levant.

Acknowledgements

This volume has been a collective effort, which was initiated from a double session Rethinking Urban Geopolitics in Ordinary and Contested Cities convened by the editors at the Royal Geographic Society (RGS) Annual Conference in London 2014.

We first and primarily want to thank our families for their limitless support and patience throughout the process of working on this book and in all our research endeavours. For Jonathan Rokem: Michal, Ben and Adam, and for Camillo Boano: Elena, Beatrice and Francesca.

We would like to thank several colleagues who have supported us throughout our work on this book manuscript, not necessarily in any specific order. We would like to thank: Julio Dávila Silva, Laura Vaughan, Caren Levy, Jennifer Robinson, Michael Safier, Matthew Gandy, James D. Sidaway, Haim Yacobi and Sara Fregonese, among several other colleagues, for their fruitful conversations and discussions at different stages of the work on this book. Thanks goes to Fok Chung Wing and Sadaf Sultan Khan for assistance with the graphics and maps. We also want to extend our thanks (especially Jonathan Rokem) to all colleagues in the Space Syntax Laboratory, Bartlett School of Architecture, UCL, and the European Commission Horizon 2020 Marie Curie Research Fellowship Grant No. 658742 for funding this project; and (especially Camillo Boano) to all colleagues in the Bartlett Development Planning Unit, UCL.

We also want to thank all the authors for their contributions, commitment and patience during the drafting process; without them this collective volume would not have come to life. Finally, we want to thank our editors at Routledge, Faye Leerink and Priscilla Corbett, for all their support during the production process.

Foreword

Sara Fregonese

> [E]ach man bears in his mind a city made only of differences, a city without figures and without form, and the individual cities fill it up.
>
> Calvino, 1978: 34

The heterogeneous corpus of literature known as urban geopolitics has, unsurprisingly for such an interdisciplinary endeavour, encountered extended critique. At least three critical strands are particularly relevant to this book in terms of how editors and contributors respond to and surpass them – advancing but also regenerating the now more than decennial agenda of urban geopolitics.

The first critiques came mainly from established political geography scholarship (Flint, 2006; Smith, 2006). These question the risk of normalizing cities as necessary *loci* of war, by simply shifting – and normalizing – the scale at which geopolitics *happens* from the national to the urban, rather than promoting a deeper understanding of urban conflict. This has not been lost on the advocates of urban geopolitics, who have warned against the peril of crystallizing knowledge about city warfare (and the violence suffered by civilians in cities) into 'a technoscientific discipline with its own conference series, research centres, and journals' (Graham, 2005: 1).

The second critique queries urban geopolitics' reliance on a handful of case studies (usually in Israel/Palestine) where militarism and warfare constitute the prism through which the urban is made sense of. Meanwhile, spaces and practices that are not a derivation of militarized conflict remain under-studied (Adey, 2013; Fregonese, 2012; Harris, 2014). This is ironic because the first mention of urban geopolitics is in Francophone scholarship within the context of power struggles in Quebec's cities (Hulbert, 1989), far from the extreme case studies that came to dominate the sub-discipline.

The third critique tackles a disembodied and techno-centric approach to urban violence: urban geopolitics tells us all there is to know about how cities can be (re)geared for war and targeted, but also hollows out these spaces of lived experiences and feeling bodies (Harker, 2014). There currently seems to be an overload of information about dramatic events of urban violence. As I write, numerous 'final messages' are being posted in real time on social media by the last residents of East Aleppo as the Syrian Army and affiliated militias move closer in December 2016. Despite this increased flux of dramatic information *from* cities at war, we somehow know still too little *about* the everyday, domestic and lived experiences and sensitivities of urban residents coping amid war, division, emergency and crisis.

This regeneration project takes stock of these critiques and tackles them not only conceptually, by bringing the everyday, the ordinary and the affective into the debate, but also by situating them within a contemporary context of global challenges, including protracted urbanized warfare and the resulting unprecedented refugee and humanitarian crises that are becoming predominantly urban.

First, this book decrystallizes urban geopolitical knowledge: it unites heterogeneous case studies and theoretical approaches under the same umbrella approach, but constantly keeps us on our toes, by seeking out the dense connections between power, space and planning at multiple scales, beyond the territorial categories of national/sub-national, foreign/domestic, state/non-state, formal/informal. The city, according to Charles Tilly, offers a 'toolbox' for researchers to link macro- and micro-scale dynamics. It is the continuous dialogue between the macro and the micro that this book mobilizes so well. One of the tenets of urban geopolitics has been *tracing connections* between macro-scale global politics and phenomena of localized violence (Graham, 2004), but this multi-scalar dialogue somehow has often faded among techno-centric analyses of mainstream urban geopolitics.

Second, the many cities, spanning four continents, in this book expand the case study range of urban geopolitics not only territorially, but also methodologically – pausing the exercise of finding similarities, and opening the ground to contrapuntal analysis and learning through distinctive differences. Taking urban geopolitics on a more intellectually refined and methodologically ambitious level, this book explores multiscale connections, along a wide range of locations, looking at the everyday and ordinary beyond the militaristic, and filling the city fabric with bodies, communities, resistances and quotidian practices.

Finally, the cities that this book takes its readers to are not only and not merely 'strategic urban networks' gaining some sort of geopolitical significance from their strength as financial and investment hubs (Sassen, in Knight Frank Research, 2012). Neither are they autonomous and bounded entities analysed solely through a specific range of technologies (Graham, 2016). Here, instead, are cities full of noises and bodies, their analysis resulting from in-depth ethnography, where the ordinary, not just the military, becomes geopolitical: 'relational sites' (Rokem and Boano, Introduction, this volume) where micro- and macro- discourses and practices are continuously reworked and contested.

Readers should not expect to contemplate these cities comfortably from afar, nor be dropped into them vertically from above. These cities 'made of differences' offer, instead, multiple access points from which to explore the ever-expanding range of conflicts, contestations and cultural formations shaping our global urban future.

References

Adey, P. (2013) Securing the volume/volumen: comments on Stuart Elden's plenary paper 'Secure the volume'. *Political Geography*, 34: 52–4. https://doi.org/10.1016/j.polgeo.2013.01.003.

Calvino, I. (1978) *Invisible Cities*, 1st Harvest/HBJ edn. New York: Harcourt Brace Jovanovich.

Flint, C. (2006) Cities, war, and terrorism: towards an urban geopolitics (book review). *Annals of the Association of American Geography*, 96(1): 216–18.

Fregonese, S. (2012) Urban geopolitics 8 years on: hybrid sovereignties, the everyday, and geographies of peace. *Geography Compass*, 6(5): 290–303. https://doi.org/10.1111/j.1749-8198.2012.00485.x.

Graham, S. (2004) Postmortem city: towards an urban geopolitics. *City*, 8(2): 165–96. https://doi.org/10.1080/1360481042000242148.

Graham, S. (2005) Remember Fallujah: demonising place, constructing atrocity. *Environment and Planning D: Society and Space*, 23(1): 1–10. https://doi.org/10.1068/d2301ed.

Graham, S. (2016) *Vertical: The City from Satellites to Bunkers*. London; New York: Verso.

Harker, C. (2014) The only way is up? Ordinary topologies of Ramallah: ordinary topologies of Ramallah. *International Journal of Urban and Regional Research*, 38(1): 318–35. https://doi.org/10.1111/1468-2427.12094.

Harris, A. (2014) Vertical urbanisms: opening up geographies of the three-dimensional city. *Progress in Human Geography*. https://doi.org/10.1177/0309132514554323.

Hulbert, F. (1989) *Essai de géopolitique urbaine et régionale: la comédie urbaine de Québec*. Laval, Quebec: Éditions du Méridien.

Knight Frank Research (2012) *The Wealth Report 2012. A Global Perspective on Prime Property and Wealth*.

Smith, N. (2006) Cities, war, and terrorism: towards an urban geopolitics (book review). *International Journal of Urban and Regional Studies*, 30(2): 469–70.

Introduction

Towards contested urban geopolitics on a global scale

Jonathan Rokem and Camillo Boano

The main focus of this collective book project is how different contested urbanisms function in diverse political contexts and geographical settings, and what we can learn from unique characteristics across different spatial social and political scales. This introduction ties the different chapter contributions together, conveying some common patterns related to planning and contestation of urban ideas and form in relation to the main overarching theme of the book: *Urban Geopolitics*. One of the growing fields of research within urban studies and political geography in the last decades is the spatio-politics of ethnically contested urban space, especially in relation to the role of planning in such sites (see, for example: Hepburn, 2004; Bollens, 2012; Allegra *et al.*, 2012). This interest should not surprise the reader, since several urbanisms and post-colonial regimes are witnessing ongoing ethnic conflicts, often violent and long-lasting. However, most of the literature published in widely influential urban circles apparently stems solely from cases in North American and European cities, with limited examples from other parts of the world (Roy, 2009; Sheppard *et al.*, 2013; Peck, 2015). This book aims to fill this gap, offering a different vision of what has been labelled the *Global South* or the *developing world* or, in other words, what has been regarded in urban studies and planning literature as a marginal *Urban Geopolitics*.

This is one of the first edited volumes covering the cross-disciplinary emerging theme of urban geopolitics from a post-colonial comparative perspective. It brings together a selected group of young and established scholars within the fields of planning, urbanism, architecture, political geography and urban sociology. Engaging with a selected group of relatively under-researched international case studies in urban studies and planning literature, it spans Latin America, East Asia, the Middle East, Africa and Europe.

The 15 local urban narratives in this volume comprise a wide range of settings: Sarajevo, Karachi, Jakarta, Khulna, Famagusta, Beirut, Jeruaslem, Acre, Cairo, Nairobi, Cape Town, Santiago, Medellín, Foz do Iguaçu and Stockholm. Each city uncovers urban geopolitics and planning at various trans-disciplinary global intersections and local scales. It is impossible to understand the history of urban politics in cities with ethnic diversity that have been at one time or another under

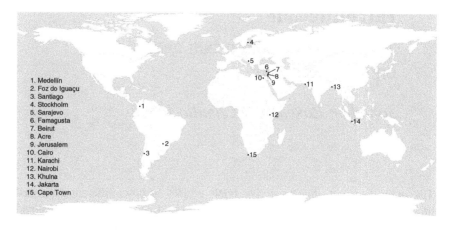

Figure 0.1 Positioning the 15 case study cities
Source: Authors, 2016

European control without relating to the colonial foundations of modern urban-ism. In this sense, colonial power relations remain an integral part of the contem-porary urban condition that still resonate spatially and geopolitically in the present (King, 1990; Jacobs, 1996; Rokem, 2016a). Critically focusing on what has traditionally been labelled as part of the 'Global South East' (Yiftachel, 2006; Watson, 2013), this edited volume's underling argument is that on the surface different kinds of contested cities share and are developing growing similarities stemming from ethnic, racial and class conflicts revolving around issues of hous-ing, infrastructure, participation and identity, among others.

The different contributions in this book consider a wide range of cities and engage with a wide range of trans-disciplinary qualitative and quantitative meth-ods. The chapters share a joint critical reading of urban geopolitics (Graham, 2004, 2010; Sidaway, 2009; Fregonese, 2009, 2012) from different urban settings with the aim of learning through differences, rather than seeking out similarities (Robinson, 2006, 2011, 2016) as part of a general call to investigate differences reframing the potential and limits of comparative urban research (McFarlane and Robinson, 2012).

In the last decade a new wave of urban research has emerged putting compari-son back on the urban studies agenda (Nijman, 2007; Ward, 2008, 2010; Robinson, 2011, 2014; Peck, 2015). However, most usual forms of comparison conventionally derive from comparing similar cases (McFarlane and Robinson, 2012) and are commonly based on a few selected cities in America and Europe (Roy, 2014), all too often focusing on the abstract city level with marginal attention given to particular local contexts (Gough, 2012). Most of this debate has been about the ontological status of the urban/city, the basis for comparing them and the consequences of different starting points in doing so

(Sidaway *et al.*, 2016: 784–5), running the risk of producing yet another wave of armchair research agenda setting lacking substantial empirical enquiry (Nijman, 2015).

With the aim of moving away from such loosely defined urban theories and contexts the book responds to McFarlane and Robinson's (2012: 766) call to place more emphasis on difference in comparative research. As such, the book questions urban studies' and planning research's long-standing Euro-centric academic knowledge production, methodological regionalism and incommensurability. To start establishing such a comparative conversation of what we can learn from different contested cities, we suggest that there is an increasing need to rethink current theoretical 'categories' and 'labels' attributed to cities, based on empirical research in a wide range of urban areas representing radically different visions and division patterns. Such a step could contribute to one of the long-standing questions at the core of urban theoretical enquiry concerning the validity of singular cases (cities) in the creation of a general urban theory (Scott and Storper, 2014).

The book's emphasis on diverse regional settings resonates with the somewhat overlooked Area Studies discipline, which has historically (mainly during the European Imperial era and later during the Cold War years) engaged with similar geographical repositionings of regional territories. Drawing on Benedict Anderson's (2016) arguments on studying intersections or making comparisons, Sidaway *et al.* (2016: 785) advocate staging comparisons in terms of four problematics: difference/similarity, expectancy/surprise, present/past and familiarity/ strangeness. These pairs are neither reducible solely to methods nor simply academic techniques, but are a discursive strategy embodying an approach of encounter and narration. Sidaway *et al.* (ibid.: 786) propose "there might be something gained by plunging into new areas, leaving a disciplinary comfort zone, with familiar literatures, paradigms and people, to think, present and publish comparatively, venturing into reconfigured area studies communities or across disciplines, where we are in less secure territory".

Our deliberation is much in line with Sidaway *et al.*'s (2016) propositions and is also infused by McFarlane *et al.*'s (2016) recent interest in intra-urban comparison (IUC), advocating new perspectives that reveal the multiple ways in which similarity and difference need to be reworked both in the context of one city and in its componentry relationality to other cities – indicating that cities are not bounded territorial containers but relational sites and processes. The comparative urbanism project should be focusing less on the city as a bond formation and more as a multiple space of many urban worlds (ibid.: 2).

Recognizing this growing interest in the world of comparative urbanisms, the chapters in this book fuse this with a myriad of contested relationships between urban space and how it structures and is structured by social life, understanding the multiplicity of urbanisms, reinforcing the need to also understand the local political, economic and social dynamics at play within urban fabrics (Boano, 2016a).

The compositional, messy, uncontrollable and recombinant nature of the present urbanism, and the differential knowledge at play in the construction of the urban, is anything but straightforward. A renewed anti essentialist shift in urban studies and practice is welcome as is 'shaking up old explanatory hierarchies and pushing aside stale concepts... making space for a much richer plurality of voices, in a way that some have likened to a democratization of urban theory'. In the critical literature, special places have been reserved for insurgent, rogue, subaltern and alt-urbanisms, as a premium has been newly attached to the disputation of generalized theory claims through disruptive or exceptional case studies.

Peck, 2015: 161

The first generative concept we wish to put forward is: *contested*, which in its very basic interpretation signifies that it contains some form of dispute, conflict and violence. Indeed, urban contestation has been taking place over centuries, in cities divided by ethnicity and race (Nightingale, 2012; Tonkiss, 2013). Presently, as a result of mounting global urban protest, there are significant debates as to the role of the welfare state, urban planning and urban space as such, in addressing the challenges of social inequalities and contested spaces in Western cities (Musterd and Ostendorf, 2013; Lloyd *et al.*, 2014: Sampson, 2013; Wacquant, 2014) and in the value of learning from other non-Western contexts (Maloutas and Fujita, 2012). Attempts to tackle stigmatized urban areas suffering from spatial and social exclusion have been well documented in the academic literature (see Marcuse and van Kempen, 2002; Andersson, 1999; Arbaci, 2007; Peach, 2009; Vaughan and Arbaci, 2011) – in other words, seeing the *urban* as "a de facto process oriented, contingent and contested condition" (Boano, 2016a: 52). However, we do wish to add an emphasis to the word *contestation*, which emerged from the Latin etymological origin of *contestari* (*litem*) from *com-* 'together' and *testari* 'to bear witness', from *testis* 'a witness', calling somehow each case and each narrative as material as witnesses as if we were in a legal combat, to contrast basic theoretical paradigmatic assumptions about the contested nature of cities.

Urban studies and planning research are associated with disagreement about theoretical and methodological issues, reflected in widely different readings across countries, cultures and contexts (Smets and Salman, 2008). One such approach, attesting to the array of different interpretations, is the growing body of literature focusing particularly on extreme 'divided' or 'contested' cities. The same cities are repeatedly cited as purportedly manifesting extreme, ethno-national divisions emanating from the 'contestation of the nation state' (Anderson, 2010; Gaffikin and Morrissey, 2011). To mention but a few, these include Belfast, Jerusalem, Johannesburg, Sarajevo, Baghdad, Beirut, Kirkuk and Mostar. These so-called 'extreme ethno-nationally divided cities' are claimed to contain distinctive attributes and tensions positioned within an exclusive category distinguishing them from other urban areas (Hepburn, 2004; Bollens, 2012; Pullan and Baillie, 2013). Moreover, within this selected group, urban transformations are analysed through Western planning perspectives, presupposing its applicability to extreme cities.

Much less attention has been given to extreme divided cities' relevance for other more peaceful cities. Following this brief review (to be further developed in each case study), we suggest that nearly all cities that contain ethnic and racial minorities as well as social and economic inequalities are *contested*, but, as the book will reveal in its wide geographical range of local cases and conclusion binding them together, *contested urbanism* may have more in common in different world regions than previously perceived in urban studies and planning literature. Hence, through a comparative approach focusing on overarching themes in each case – such as housing, infrastructure, participation and identity, among others (Rokem, 2016a) – we aimed to illustrate how cities are geopolitical spaces as they are embedded in a web of contested visions where the production of space is an inherently conflictive process, manifesting, producing and reproducing various forms of injustice, as well as alternative forces of transgression and social projects (Boano *et al.*, 2013). At the same time, we aim to expose the opportunities and challenges faced by a growing post-colonial understanding (Robinson, 2006; Edensor and Jayne, 2011; Oldfield and Parnell, 2014; Roy, 2016) of planning, urban development and urban geopolitics from a broader global contested urbanism comparative perspective.

However, we not only suggest a new urban ontology that moves beyond the 'West' or 'North', but also wish to add a comparative and relational understanding of the contested nature that the so-called *southern cities* are developing – on the one hand, as a result of neoliberalism and growing inequalities and, on the other hand, a surge in ethnic identity politics and nationalism. In doing so, the book is repositioning contestation at the centre of urban research, addressing the intersection of spatial and temporal aspects of conflicts in the production of the city, where intellectual and spatial categories are able to construct new epistemologies positioning cities and space in a paradoxical tension (Boano, 2016a)

Moreover, this book adds a renewed urban geopolitical dimension pointing at the need to rethink current 'categories' and 'labels' and, as such, critically questioning the enduring 'North-Western'/'South-Eastern' divide within urban studies and planning research. The book offers an in-depth understanding of the worldwide contested nature of cities with a detailed review from a wide range of local contexts peripheral yet pertinent to universalizing urban studies and planning theory beyond the prevailing Euro-American debate.

The book, through its diverse geographical foci, suggests that it is time to set a new research agenda to regenerate the emerging sub-field of urban geopolitics bridging the disciplines of political geography, urban studies and planning. Recent adaptations to classic geopolitics (Dalby, 1990; Agnew, 2003) have seen an increasing interest in placing cities beyond the usual geopolitical focus of state power and territorial control, scaling down to local sites shaping what we frame here as an emerging 'urban geopolitical turn' (Graham, 2004, 2010; Sidaway, 2009; Fregonese, 2009, 2012; Yacobi, 2009). Urban geopolitics has traditionally stemmed from two main bodies of research, both using it differently as a synonym for an urban political geography in an age of terror. First, engaging with the militarization of urban space, surveillance and security (Graham, 2004, 2010; Gregory, 2011) and asymmetric vertical urban warfare (Weizman, 2007), this has led to a deeper scrutiny of cities and their containment of material damage and targeted violence. Second, in the past two decades, a fast-evolving strand within urban political geography and planning has focused on urban conflicts within ethno-nationally contested cities, especially in relation to the role of planning (Hepburn, 2004; Anderson, 2010; Bollens, 2012; Rokem, 2016a).

In an era of growing neoliberalization, ethno-nationalism and international migration, there is a growing need to critically examine urban geopolitics as significant lens to encapsulate recent shifts in the global urban present. Violence, disaster and division can no longer be ignored in a century where the majority of the world population is urban (Fregonese, 2012: 298). In this context, Newman (2006) proposes that the impact of borders and territoriality is not diminishing; rather, new scales of territorial affiliations and borders are recognizable that may be flexible but are still selective on different geographical scales. In other words, while, traditionally, the national affiliation of cities has tended to be attached to the nation state, the question now arises if this still remains the case. This is echoed by rising claims for urban recognition and sovereignty from a growing number of immigrants and refugees living in cities and camps far away from their original homeland. In this process cities are reshaping both spatially and socially, creating new forms of *urban geopolitical* actors and scales across distant national and cultural conflicts pivoted at the urban scale.

Specifically important for us and for the cases we selected in the book is that conflicts and political violence alike have not only direct spatial implication visible to all in the form of destruction, seclusion and control, but also unfold at various interconnected scales: global, territorial, state, urban, human. Their geographical scopes stretch from the localized sites of citizen contestation and micro-struggles to the global networks of terror, with different modes of visibility and intelligibility. Conflicts transform land uses, territorial arrangements, urban processes and human settlement patterns according to temporalities that range from short-lived states of emergency to the *longue durée* of chronic violence, permanent occupations and predatory urbanisms (Boano, 2016a). The present international geopolitical conditions with large-scale forced migration and lack of local integration is having a substantial impact on the urban geopolitical condition, with relatively limited attention given to the 'planning politics nexus': the relation between planning and politics, as a non-hierarchical set of interactions, negotiated within the specific historical, geographical, legal and cultural contexts (Rokem and Allegra, 2016) and its impact across different supranational, national and local scales.

The very question that remains open is whether one should challenge the canonical differentiation between causal categories of spatial segregation, division and conflict (i.e. driven by market gentrification, state led or social dynamics, with the latter perhaps encompassing some form of societal *othering* of individuals and communities) (Rokem, 2016b: 406). We suggest there is a need to move beyond the focus on cities as direct targets of terror and violence by different state and non-state actors. As such we need to re-engage in a critical reading of different contestation patterns in cities and towards a closer assessment of political geography with a new understating of the *post-colonial, ordinary, domestic, embodied* and *vertical dimensions* to better comprehend recent shifts in urban geopolitics thinking.

The book brings together a selected group of international empirically grounded cases engaging with urban planning and geopolitics from a set of different cities worldwide. In doing so, this edited volume seeks to argue that it is timely to start learning from, and compare across, different urban case studies (Abu Lughod, 2007) utilizing IUCs (McFarlane *et al.*, 2016) and advocating staging comparisons in terms of problematics (Sidaway *et al.*, 2016) exposing one urban context's relational and contrastive relevance to other cities (Rokem, 2016a), suggesting there is a growing need to rethink 'labels' and 'concepts' attributed to cities and neighbourhoods, to better conceptualize and adapt policy and practice to ethnic minorities and migrants in an ever more fractured urban geopolitical present. In so doing, we question what we can learn from clutching the universal complexities of different contestation patterns in cities not traditionally part of the dominant theory-building cases – to advance our understanding of urban studies, development studies and planning in the twenty-first-century contested urban reality.

The volume is structured across five parts. The Foreword by Sara Fregonese, concluding conversation with Michal Safier and Afterword by James D. Sidaway frame some of the central past, present and future lineages of the urban geopolitical debate. The three chapters in the opening part, 'Comparative urban geopolitics', engage with a relational and contrastive conceptualization of the value of urban comparisons learning from: Sarajevo and Beirut; Nairobi and Cape Town; and Stockholm and Jerusalem. This opening part operates dual urban comparisons, setting the tone for the next three regional parts, all with a more particular focus on cities from the Far East (Karachi, Jakarta and Khulna), Middle East (Famagusta, Acre and Cairo) and Latin America (Santiago, Medellín and Foz do Iguaçu). While authors have diverse trans-disciplinary backgrounds spanning geography, planning, architecture, development studies and urban sociology, among others, the 12 chapters relate to local empirical manifestations of *contested urbanisms*, employing different methodological techniques and theoretical frameworks. The overall aim is to advance the cross-disciplinary field of urban geopolitics, bringing geopolitics into the mainstream agenda of urban studies, to enhance our understanding of cities as contested nexus points of social, spatial and political change across different geographical scales.

In the opening chapter of the volume, Gruia Bădescu provides an historical comparative interpretation of Sarajevo and Beirut's shared antagonistic urban imaginaries of cosmopolitanism and contestation, as well as experience of urban warfare, segregation and post-war reconstruction. Scrutinizing how, despite these similarities, the process of urban reconstruction after the Lebanese Civil War (1975–90) and the war in Bosnia and Herzegovina (1992–5), respectively, produced very different outcomes resulting from contrasting post-war planning frameworks and political settlements, Bădescu, argues that, in these particular contexts, the city does not emerge as an autonomous body circumventing national politics – becoming, instead, an arena of conflicting urban geopolitical articulations of state-level politics and local ethnic and religious dynamics.

The second chapter moves us to a different regional focus, on two of the fast-growing African metropolises comparing Nairobi and Cape Town. Liza Rose Cirolia explains, in her chapter, that in many large African cities there has been no central entity effectively controlling development or upholding a 'public mandate' to invest in infrastructure. Non-delivery of infrastructure has become the norm rather than the exception. Cirolia maintains Nairobi's urban development story highlights the multi-dimensional nature of informality in the city. While, in Cape Town, there is a shift from apartheid planning that included the creation of zones for white, coloured and black African households to a more neoliberal-planning regime, more generally the chapter calls for a need to move beyond the continued focus on racial segregation that still dominates the South African cities' academic urban planning literature.

In the third chapter, Jonathan Rokem closes the first comparative part with a contrastive and relational assessment of ethnic minority segregation in two radically different ethnically contested cities. The central proposition put forward is that there is much to learn from a comparative investigation of spatial and social policies towards ethnic minorities in Jerusalem and Stockholm. The chapter facilitates a multi-scalar comparative conversation of urban difference. Considering three cross-cutting themes: (1) housing and development, (2) mobility and transport and (3) local government and civil society, Rokem suggests that it is timely to start comparing across different ethnically contested cities as part of a general call to rethink our understanding of incommensurable cases in the contemporary urban research and practice.

Opening the second part, covering three South and South East Asian cities, Sadaf Sultan Khan, Kayvan Karimi and Laura Vaughan portray an illumining in-depth spatial investigation of the changing political fortunes of the *Muhajir* community, Karachi's largest migrant group. Karachi, the capital of Pakistan, is well known for its violent ethnic and sectarian conflict, with different communities striving for dominance. This chapter aims to articulate the synergistic relationship between the city's urban planning strategies and the complex ethno-politics of its many migrant communities. The authors suggest that what appears to have happened in Karachi over the course of the last half-century is a process of inversion of power, where a national minority has been able to control and transform the districts in which it constitutes a majority group. This local urban geopolitical dominance has been achieved through the combined impact of legitimate political processes and violent street presence. Next, Pawda F. Tjoa explores the spatial politics of contested urban space through the lens of 'marketplace coordination' in Jakarta by highlighting the roots of social tension and the escalation of internal conflict during the period 1997–8. Tjoa reveals how urban policies geared towards creating order and progress triggered permutations of social categories within local merchant communities. Conflicts between market stalls created social tensions that erupted during the Asian financial crisis. Thus, the ideology of 'development' became a catalyst for conflict, which contributed to the persistence of urban geopolitical fragmentation. Apurba Kumar Podder closes the second part with a critical and radical rethinking of one aspect of poverty culture, commonly seen as 'doing nothing'. While the academic scholarship often explains doing nothing as idleness or a response of the poor to societal alienation, Podder offers a novel perspective exploring a case of an illegal bazaar located in Khulna, one of the southern cities in Bangladesh. Through a local ethnographic exploration, he argues that doing nothing should be understood as an alternative mode of the poor's occupational urban geopolitics to sustain in a condition of unequal power relations.

In the opening chapter of the third part, focusing on cities from the Middle East and North Africa, Moriel Ram exposes theoretical, thought-provoking formation processes of urban 'spaces of exception', giving the example of the city of Famagusta, in Northern Cyprus, which has been under Turkish military occupation since 1974. Ram argues that conquest and ensuing Turkish occupation constantly produces an urban threshold between Famagusta and two competing spatial processes of encampment, exclusion and seclusion: the enclosed area of Varosha and the campus of Eastern Mediterranean University. The chapter demonstrates how the threshold between Varosha and Famagusta provides for urban geopolitical legitimacy of the urban space, while the link between the campus and the city economically sustain the latter.

The second chapter focuses on the mixed northern Israeli city of Acre. Nimrod Luz and Nurit Stadler examine the complex and reflexive relations between urbanity, religion and ethnicity, suggesting there is much to learn from minority religious claims and religious spatialities and the challenges of these claims by hegemonically opposing groups. Focusing on a local urban struggle revolving around the reconstruction of a mosque by the Muslim minority, Luz and Stadler critically reflect on the city's transformation in terms of its religious voices, planning processes, everyday life and contested urban geopolitics.

Cairo concludes the third part. The city has held a central position as a pivotal intersection of Africa, Asia and Europe. Mohamed Saleh argues that, for decades, public space in Egypt has been systematically deprived of its essential symbolic functions. Upon integrating the country into the global model of neoliberalism, the state has adopted public policies on various scales, which resulted in a deep-rooted crisis of participation and identity. Saleh utilizes the notions of complexity to explore the roots of this crisis, perceiving the results as a path-dependent structural shift in society, stemming from thresholds that stretch back from the post-colonial condition to the rise of neoliberalism and social media.

The fourth part of the book offers an overview of three Latin American cities, each with distinct urban processes showcasing the deep social and spatial inequality and the need to diversify the understanding of local urban geopolitics. Camila Cociña and Ernesto López-Morales explore the role that gentrification processes can potentially have in the emergence of non-violent conflict, allowing local organizations to participate in the encounter of clearly differentiated positions in Santiago, Chile. Cociña and López-Morales discuss three local urban cases in which spaces of conflict have allowed the development of alternatives whereby less affluent groups manage to remain in gentrifying neighbourhoods. The authors argue that exploring the notion of conflict as an essential part of democracy during class encounters in gentrification processes can be seen as an opportunity to shift urban geopolitical power struggles through local mobilization of active community groups.

Next, Catalina Ortiz and Camillo Boano put forward a more critical scrutiny of Medellín's aspiration of consolidating as a global benchmark of urban innovation. Urging for the development of new ways of thinking on how this model enables (or inhibits) opportunities for spatial justice, Ortiz and Boano reflect on the production and rearrangement of urban space driven by an urban geopolitics of informality and the politics of securitization and control. Focusing specifically on Comuna 8, in the central-east area of Medellín, and everyday citizens' politics in informal settlements, it is argued that the politics of informality operate as a governmental technology that strategically uses the denial, self-provision or monumentalization of infrastructure as a means of selectively legitimizing or criminalizing citizens' claims over space.

In the final chapter of the fourth part, Peter D. A. Wood offers a methodologically distinctive perspective on how to measure trans-border participation through urban development planning in Foz do Iguaçu, Paraná, Brazil. Wood suggests that, through use of a Q methodology, opinions on who participates in development, particularly within this Brazilian borderland city, can be revealed. These results are then used to establish three key worldviews among those involved with or affected by development planning in the region: local integration optimists, institution sceptics and nationalists. Through further examination of the collected data, the author calls for a more critical attitude and geopolitical approach to urban studies.

In the concluding conversation held throughout a series of meetings in London in 2015–16, Jonathan Rokem and Camillo Boano reflect with Michael Safier, who dedicated most of his research to conceptualizing a *cosmopolitan development* framework as a means of promoting peaceful coexistence and dialogue between cultural groups in urban areas. Starting in the early 1990s, Safier proposed *cosmopolitan urbanization* as a way forward to capture all the varieties of interaction between different cultural traditions, heritages, identities and practices in urban life (Safier, 1993). In this concluding conversation we discuss some of Safier's central ideas, which are much in line with what this collective book project aims to offer within the emerging research field of *comparative urban geopolitics*. We especially consider connotations concerning emerging threats from varied and destabilizing combinations of global, regional and local urban inequalities. Safier asserts there are two ways to respond to this danger: one based on withdrawal, underlined by exclusionist, intolerant and even aggressive reaction; and the other one based on active engagement, borne by inclusion and coexistence, whereby cities would be central arenas in which these conflicts and reconciliations cumulate.

References

Abu Lughod, J. (2007) The challenge of comparative case studies. *City*, 11(3): 399–404.

Agnew, J. (2003) *Geopolitics: Re-visioning World Politics*, 2nd edn. London: Routledge.

Allegra, M., Casaglia, A., and Rokem, J. (2012) The political geographies of urban polarization: a review of critical research on divided cities. *Geography Compass: Urban Section*, 6(9): 560–74.

Anderson, B. (2016) Frameworks of comparison. *London Review of Books*, 38(2): 15–18.

Anderson, J. (2010) *Democracy, Territoriality and Ethno-National Conflict: A Framework for Studying Ethno-Nationally Divided Cities*. Paper no.18. www.conflictincities.org. Accessed May 2015.

Andersson, R. (1999) 'Divided Cities' as a policy-based notion in Sweden. *Housing Studies*, 14(5): 601–24.

Arbaci, S. (2007) Ethnic segregation, housing systems and welfare regimes in Europe. *European Journal of Housing Policy*, 7(4): 401–33.

Boano, C. (2016a) Jerusalem as a paradigm: Agamben's 'whatever urbanism' to rescue urban exceptionalism. *City*, 20(3): 419–35.

Boano, C. (2016b) La ciudad imposible: breves reflexiones sobre urbanismo, arquitectura y violencia. *Materia Arquitectura*, 12: 49–57.

Boano, C., Hunter, W., and Newton, C. (2013) *Contested Urbanism in Dharavi: Writings and Projects for the Resilient City*. London: The Bartlett Development Planning Unit.

Bollens, S. A. (2012) *City and Soul in Divided Societies*. London and New York: Routledge.

Dalby, S. (1990) American security discourses: the persistence of geopolitics. *Political Geography Quarterly*, 9: 171–88.

Dalby, S. (2010) Recontextualising violence, power and nature: the next twenty years of critical geopolitics? *Political Geography*, 29(5): 280–8.

Edensor, T., and Jayne, M. (2011) *Urban Theory Beyond the West: A World of Cities*. London: Routledge.

Fregonese, S. (2009) The urbicide of Beirut? Geopolitics and the built environment in the Lebanese civil war (1975–1976). *Political Geography*, 28(5): 309–18.

Fregonese, S. (2012) Urban geopolitics 8 years on: hybrid sovereignties, the everyday, and geographies of peace. *Geography Compass*, 6(5): 290–303.

Gaffikin, F., and Morrisey, M. (2011) *Planning in Divided Cities*. New Jersey: Blackwell.

Gough, K. V. (2012) Reflections on conducting urban comparisons. *Urban Geography*, 33(6): 866–78.

Graham, S. (ed.) (2004) *Cities, Wars and Terrorism, Towards an Urban Geopolitics*. New Jersey: Blackwell.

Graham, S. (2010) *Cities Under Siege: The New Military Urbanism*. London, New York: Verso.

Gregory, D. (2011) Lines of descent. *Open Democracy*. www.opendemocracy.net/derek-gregory/lines-of-descent. Accessed September 2016.

Hepburn, A. C. (2004) *Contested Cities in the Modern West*. New York: Palgrave.

Jacobs, M. J. (1996) *Edge of Empire Post Colonialism and the City*. London: Routledge.

King, A. (1990) *Urbanism, Colonialism and the World Economy*. London: Routledge.

Lloyd, C., Shuttleworth, I., and Won, D. (2014) *Social-Spatial Segregation*. London: Policy Press.

Maloutas, T., and Fujita, K. (2012) *Residential Segregation in Comparative Perspective*. Farnham: Ashgate.

Marcuse, P., and van Kempen, R. (eds) (2002) *Of States and Cities: The Partitioning of Urban Space.* Oxford and New York: Oxford University Press.

McFarlane, C., and Robinson, J. (2012) Introduction: experiments in comparative urbanism, *Urban Geography*, 33(6): 765–73.

McFarlane, C., Silver, J., and Truelove, Y. (2016) Cities within cities: intra-urban comparison of infrastructure in Mumbai, Delhi and Cape Town. *Urban Geography*, doi: 10.1080/02723638.2016.1243386.

Musterd, S., and Ostendorf, W. (2013) *Urban Segregation and the Welfare State: Inequality and Exclusion in Western Cities.* London: Routledge.

Newman, D. (2006) The lines that continue to separate us: borders in our borderless world. *Progress in Human Geography*, 30(2): 1–19.

Nightingale, C. H. (2012) *Segregation: A Global History of Divided Cities.* Chicago: University of Chicago Press.

Nijman, J. (2007) Introduction: comparative urbanism. *Urban Geography*, 28(1): 1–6.

Nijman, J. (2015) The theoretical imperative of comparative urbanism: a commentary on 'Cities beyond compare?' by Jamie Peck. *Regional Studies*, 49(1): 183–186.

Oldfield, S., and Parnell, S. (2014) *Handbook on Cities in the Global South.* London: Routledge.

Peach, C. (2009) Slippery segregation: discovering or manufacturing ghettos? *Journal of Ethnic and Migration Studies*, 35(9): 1381–95.

Peck, J. (2015) Cities beyond compare? *Regional Studies*, 49(1): 183–6.

Pullan, W., and Baillie, B. (eds) (2013) *Locating Urban Conflicts: Ethnicity, Nationalism and the Everyday.* London: Palgrave Macmillan.

Robinson, J. (2006) *Ordinary Cities: Between Modernity and Development.* London and New York: Routledge.

Robinson, J. (2011) Cities in a world of cities: the comparative gesture. *International Journal of Urban and Regional Research*, 35(1): 1–23.

Robinson, J. (2014) Introduction to a virtual issue on comparative urbanism. *International Journal of Urban and Regional Research.* Available from: http://onlinelibrary.wiley.com/doi/10.1111/1468-2427.12171/full. Accessed August 2015.

Robinson, J. (2016) Thinking cities through elsewhere: comparative tactics for a more global urban studies. *Progress in Human Geography*, 40(1): 3–29.

Rokem, J. (2016a) Beyond incommensurability: Jerusalem and Stockholm from an ordinary cities perspective. *CITY*, 20(3): 451–61.

Rokem, J. (2016b) Introduction: learning from Jerusalem – rethinking urban conflicts in the 21st century. *CITY*, 20(3): 472–82.

Rokem, J., and Allegra, M. (2016) Planning in turbulent times: exploring planners agency in Jerusalem. *International Journal of Urban and Regional Research.* doi:10.1111/1468-2427.12379.

Roy, A. (2009) The 21st-century metropolis: new geographies of theory. *Regional Studies*, 43(6): 819–30.

Roy, A. (2014) Before theory: in memory of Janet Abu-Lughod. Available from: www.jadaliyya.com/pages/index/16265/before-theory_in-memory-of-janet-abu-lughod. Accessed March 2014.

Roy, A. (2016) Who's afraid of postcolonial theory? *International Journal of Urban and Regional Research*, 40(1): 200–9.

Safier, M. (1993) The case for cosmopolitan studies: understanding the cultural dimensions of urban development. Memorandum. Unpublished.

Sampson, R. J. (2013) *Great American City: Chicago and the Enduring Neighbourhood Effect.* Chicago: Chicago University Press.

Scott, A. J. and Storper, M. (2014) The nature of cities: the scope and limits of urban theory. *International Journal of Urban and Regional Research*, 39(1): 1–16.

Sheppard, E., Leitner, H., and Maringanti, A. (2013) Provincializing global urbanism: a manifesto. *Urban Geography*, 34(7): 893–900.

Sidaway, J. D. (2009) Shadows on the path: negotiating geopolitics on an urban section of Britain's South West Coast Path. *Environment and Planning D*, 27(6): 1091–116.

Sidaway, J. D., Ho, E. L. E., Rigg, J. D., and Woon, C. Y. (2016) Area studies and geography: trajectories and manifesto. *Environment and Planning D: Society and Space*, 34(5): 777–90.

Smets, P., and Salman, T. (2008) Countering urban segregation: theoretical and policy innovations from around the globe. *Urban Studies*, 45(7): 1307–32.

Tonkiss, F. (2013) *Cities by Design: The Social Life of Urban Form*. London: Polity Press.

Vaughan, L., and Arbaci, S. (2011). The challenges of understanding urban segregation. *Built Environment*, 37(2): 128–38.

Wacquant, L. (2014) Marginality, ethnicity and penalty in the neoliberal city: an analytic cartography. *Ethnic and Racial Studies*, 37(10): 1687–711.

Ward, K. (2008) Editorial: toward a comparative (re)turn in urban studies? Some reflections. *Urban Geography*, 29(5): 405–10.

Ward, K. (2010) Towards a relational comparative approach to the study of cities. *Progress in Human Geography*, 34(4): 471–87.

Watson, V. (2013) Planning and the 'stubborn realities' of global south-east cities: some emerging ideas. *Planning Theory*, 12(1): 81–100.

Weizman, E. (2007) *Hollow Land*. London: Verso.

Yacobi, H. (2009) Towards urban geopolitics. *Geopolitics*, 14(3): 576–81.

Yiftachel, O. (2006) Ethnocracy. MA, University of Pennsylvania Press.

Part I

Comparative urban geopolitics

Jonathan Rokem

In recent years a new wave of comparative research has emerged putting urban comparisons back on the agenda (Nijman, 2007; Ward, 2008, 2010; McFarlane, 2010; Robinson, 2006, 2011, 2016; Peck, 2015; Sidaway *et al.*, 2016; McFarlane *et al.*, 2016). However, most usual forms of comparison conventionally depend on associating between cities with similar geo-historical settings (McFarlane and Robinson, 2012) and are commonly based on a few selected cities in North America and Europe (Roy, 2009; Peck, 2015).

With the aim of moving away from comparing similar cases in Euro-America, the three chapters in this first part of the book cover a wide range of urban geopolitical processes in different regional settings and within diverse theoretical discourses, with the aim of placing more emphasis on difference in comparative urban research. As such, the overall conceptual structure of this first part is an introductory lens for the whole book, questioning the position of post-colonial theory within urban studies' enduring Euro-centric academic knowledge production, methodological regionalism and incommensurability. Each of the three chapters establishes a comparative conversation of what we can learn from different urban contexts. In the first chapter, Gruia Bădescu compares post-war reconstruction in two contested cities: Sarajevo and Beirut. Liza Rose Cirolia explores planning innovations in two African cities, Nairobi and Cape Town, in the second chapter. And, in the third chapter, Jonathan Rokem contrasts urban ethnic segregation in different contested cities: Jerusalem and Stockholm. The authors of the three chapters suggest different ways to rethink current theoretical categories and labels, based on empirical comparative research in two cities, representing both relational and contrastive – yet radically different – visions and division patterns.

References

McFarlane, C. (2010) The comparative city: knowledge, learning, urbanism. *International Journal of Urban and Regional Research*, 34(4): 725–42.
McFarlane, C., and Robinson, J. (2012) Introduction: experiments in comparative urbanism, *Urban Geography*, 33(6): 765–73.

McFarlane, C., Silver, J., and Truelove, Y. (2016) Cities within cities: intra-urban comparison of infrastructure in Mumbai, Delhi and Cape Town, *Urban Geography*, doi: 10.1080/02723638.2016.1243386.

Nijman, J. (2007) Introduction: comparative urbanism. *Urban Geography*, 28(1): 1–6.

Peck, J. (2015) Cities beyond compare? *Regional Studies*, 49(1): 183–6.

Robinson, J. (2006) *Ordinary Cities: Between Modernity and Development*. London and New York: Routledge.

Robinson, J. (2011) Cities in a world of cities: the comparative gesture. *International Journal of Urban and Regional Research*, 35(1): 1–23.

Robinson, J. (2016) Thinking cities through elsewhere: comparative tactics for a more global urban studies. *Progress in Human Geography*, 40(1): 3–29.

Roy, A. (2009) The 21st-century metropolis: new geographies of theory. *Regional Studies*, 43(6): 819–30.

Sidaway, J. D., Ho, E. L. E., Rigg, J. D., and Woon, C. Y. (2016) Area studies and geography: trajectories and manifesto. *Environment and Planning D: Society and Space*, 34(5): 777–90.

Ward, K. (2008) Editorial: toward a comparative (re)turn in urban studies? Some reflections. *Urban Geography*, 29(5): 405–10.

Ward, K. (2010) Towards a relational comparative approach to the study of cities. *Progress in Human Geography*, 34(4): 471–87.

1 Post-war reconstruction in contested cities

Comparing urban outcomes in Sarajevo and Beirut

Gruia Bădescu

Introduction

Cities of the Balkans and the Middle East have been traditionally discussed in relationship to diversity, exchange and cosmopolitanism, but also to conflict and urban violence (Ilbert, 1996; Mazower, 2005; Driessen, 2005; Freitag and Lafi, 2014; Freitag *et al.*, 2015). An increasing recent focus on studying urban conflicts has prompted the analysis of several cities in the region under the umbrella of 'divided city in contested states' research, ranging from Nicosia to Jerusalem, and either pointing out their planning conundrums (Bollens, 2006, 2012; Pullan, 2013) or exploring and situating everyday practices in such contested urban space (Bakshi, 2011, 2014; Pullan and Baillie, 2013; Pullan *et al.*, 2013). Beirut and Sarajevo – contextualized and framed differently in area studies research, but brought together by the divided and contested cities work – have been highlighted for showcasing the epitome of urban conflicts: war in the city, or, according to Bogdanović (1993) and Coward (2006, 2009), urbicide – a war on the city, urban space and urbanity. 'Urbicide', as a key new term in urban geopolitics (Graham, 2004; Fregonese, 2009) led to close investigations of meanings of destruction of urban environment in conflict. Through their wartime experience, Beirut – scene of the Lebanese Civil War (1975–90) and aerial attacks by Israel in 2006 – and Sarajevo under siege as part of the war in Bosnia and Herzegovina (1992–5) – were shaped as urban geopolitical spaces. As Fregonese (2009: 317) showed in her study of wartime Beirut, through various practices, including assault and destruction, possession and partition, Beirut's urban environment became geopolitical, as "it was the product and the tool of the re-territorialisation of the city". However, as Brand and Fregonese point out (2013), most literature has focused on the war period itself; they call for an investigation of processes occurring in the aftermath. Post-war reconstruction is such a process that provides a lens to rethink planning and urban geopolitics in contested cities, as well as their articulations in everyday practices. This chapter examines the process of urban-architectural post-war reconstruction in Beirut and Sarajevo in an urban geopolitical frame. It explores how urban environment both responds to and shapes territorialization processes and how the discussion of urban geopolitics and the relationship between urban space and the state can be enriched by urban comparisons.

Beirut and Sarajevo are particularly fit for a comparison. The two cities share an urban history including a long period of Ottoman rule followed by the protectorate of a European power (the Habsburg state and France, respectively). After the Second World War, the two cities witnessed a significant expansion and embracement of architectural modernism in new state frameworks (socialist Yugoslavia and independent Lebanon). Beirut had a reputation as the Paris of the Middle East, a sophisticated, thriving city at the edge of the Mediterranean (Sawalha, 2010), while Sarajevo was a celebrated symbol of urban coexistence, especially after the 1984 Winter Olympics, which portrayed it to the world as a confident place of harmonious relations between a diverse population. In fact, both cities were hailed as examples of harmonious, pluralistic urban societies (Donia, 2006; Saliba, 1998). The populations of both cities included a mix of Christians and Muslims who shared a common language – Arabic in Beirut and what was called, for most of the twentieth century, Serbo-Croatian in Sarajevo. The urban imaginaries of harmonious coexistence between the groups were challenged by occasional outbursts of tension and violence. These mirrored the contrasting historiographies of both Lebanon and Bosnia and Herzegovina, which included opposed accounts of pluralism and peaceful coexistence versus antagonism and conflict as main lenses of these societies at large (Salibi, 1988; Donia, 1994; Naimark and Case, 2003; Ramet, 2006). Both this diversity and the related contrasting interpretative frames were also expressed in the urban landscape. Their shared Ottoman history accounted for the spatial structure defined by the *mahalas*, residential neighbourhoods centred on one house of worship, with members of the respective religious communities clustering in that neighbourhood, and a commercial core (Donia, 2006; Salibi, 1988). The clustering of religious buildings in the city centres of Beirut and Sarajevo, representing all of the two cities' faiths (Figure 1.1), has been usually interpreted as an indication of harmonious cohabitation and pluralism. Nevertheless, this was challenged by Hayden (2013), who sees the spatial proximity as an expression of a thin balance and minimal tolerance, not of harmony and pluralism. He defines the clustering as antagonistic 'religioscapes', which would contribute to the contestation of space through the coexistence of multiple claims in adjacent space. This dichotomy, harmony–antagonism, has marked long-standing contrasting understandings of Bosnian and Lebanese urban space, echoing the general discussion on cities of the eastern Mediterranean (Freitag *et al.*, 2015).

During the Lebanese Civil War and the Bosnian War, Beirut and Sarajevo became poster images of war and widespread destruction. Beirut was the scene of infighting between Christian and Muslim paramilitaries, Palestinian groups, Israeli and Syrian armies for almost 15 years (1975–89). The siege of Sarajevo – the longest siege of a capital city in the history of modern warfare – lasted 1,395 days, between 1992 and 1996; the Bosnian Serb paramilitary and the Yugoslav National Army hit the city with an average of 329 shells per day (Potyrala, 2009). War led to significant destruction, which in an urban geopolitics framework could be understood as more than collateral damage, including targets related to the identity of particular groups (i.e. religious buildings) (Bevan, 2007), or of entire

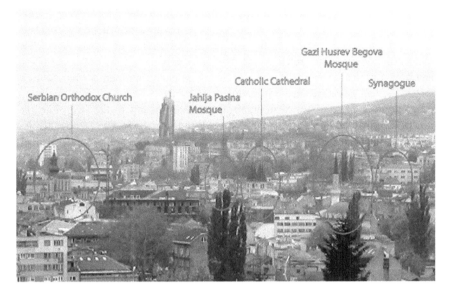

Figure 1.1 Religious buildings for four different faiths lie close to each other in downtown Sarajevo

Source: Author, 2010

urban infrastructures and institutions (Graham, 2004) – as a war against common urban life, described as urbicide (Bogdanović, 1993; Coward, 2009). Furthermore, war led to a segregation of residential geographies along ethno-religious lines that remains to date. Beirut and Sarajevo thus came to exemplify ethno-religiously contested cities, witnessing division and destruction, followed by challenging processes of urban geopolitical reconstruction.

Reconstruction is not only defined by its temporality – the period after war – but it also encompasses a process and a set of policies that include, according to Barakat (2005: 11), "a range of holistic activities in an integrated process designed not only to reactivate economic and social development but at the same time to create a peaceful environment that will prevent a relapse into violence". Reconstruction has been seen by most practitioners and many researchers, first, as a functional process (Barakat, 2005). The process of post-war urban reconstruction is highly complex, with the concerns of public authorities, urban planners and architects ranging from housing provision to infrastructure, from what should be rebuilt in its past form to what should be rethought as new. In its systemic, integrated approach and its urban outcomes, one can see similarities with the processes of urban regeneration and urban reconfiguration that affect any 'ordinary city'. Nevertheless, reconstruction is also a symbolic, political process in which choices made could be shaped by understandings of the past or of the desired future. The relationship between reconstruction and socio-political processes of reconciliation, as well as the positionality of city-makers, has been touched upon in a number of studies (Barakat *et al.*, 2008; Bollens, 2006; Calame and Charlesworth, 2009; Lyons *et al.*, 2010). This chapter enquires how post-war reconstruction in contested cities prolongs and transforms the nature of contestation of urban space; how reconstruction outcomes relate to the politically expressed forms of contestation; and how a critical urban geopolitics, focusing on analysing memory narratives as discourse, can advance our understandings of contested cities. It examines how, despite similar patterns in demography and urban history in the two cities, the process of post-war urban reconstruction produced very distinctive urban outcomes in Beirut and Sarajevo. It explores how public authorities, architects and planners used post-war reconstruction as a way not only to repair the urban fabric, but also to express spatially different constellations of memory and contestation by various groups. It argues that, in these particular contexts, the city does not emerge as an autonomous body that can circumvent national politics, but becomes instead both an arena of state-level politics and local dynamics.

Post-war state political arrangements as a frame for urban reconstruction

Throughout the Lebanese Civil War and the war in Bosnia and Herzegovina, architects and planners were already debating the ways reconstruction would take place (Jamaković and Association of Architects, 1993; Ragette, 1983). In the words of Esther Charlesworth, Beirut became "the world's largest laboratory for post-war reconstruction" (Charlesworth, 2006: 54). Local and international architects and planners held meetings throughout the 1980s, aiming to draw lessons from reconstruction in Europe after the Second World War, but also to think through the specificities of dealing with an increasingly divided, polycentric city (Ragette, 1983). Similar debates occurred in Sarajevo. Local Sarajevo project *Warchitecture* tried to make sense of destruction and start a discussion on the future of the city (Jamaković and Association of Architects, 1993). In both cities, there were similar professional debates concerning how reconstruction can be used as an opportunity for functional improvement in city flows, infrastructure and transport. Yet, for some, it was a call for rethinking cities, architecture and the autonomy of urban communities. "War, disaster gives us the capacity to rethink architecture," wrote American architect Lebbeus Woods (1997: 1) when he visited Sarajevo in 1993, working with local architects on realizing his concept of radical reconstruction, in which, beyond the functional reconstruction of the built environment, the scars of buildings are articulated in the creation of free spaces to be used by the community. All in all, functional, radical or symbolic approaches worked within assumptions of certain autonomy of the city in reconstruction decision-making.

None of these debates were, however, translated into practice until the peace agreements were reached and a new political framework was shaped. Despite the many similarities pointed out, the post-war trajectories of Beirut and Sarajevo were very different, occurring in two distinguishing geopolitical arrangements.

The *Ta'if* Agreement of October 1989 for Lebanon and the Dayton Accords of November 1995 provided the political framework for the post-war reconstruction processes in the two countries and their respective capital cities. The *Ta'if* Agreement was signed in Saudi Arabia just before the fall of the Berlin Wall, the product of negotiations mediated by Saudi Arabia and the USA, with Syria and other Arab states in the background, with the intent of not only ending the war, but also reintegrating Lebanon in the Arab world and reasserting its Arab identity (Hudson, 1997; Salamey, 2013). Dayton, brokered by a USA frustrated by European non-action, occurred in a post-Cold War era in which the USA emerged as a sole superpower which sees itself to have a – selective – duty to prevent war, intervene to stop conflict (Gulf War, Somalia), or create diplomatic frameworks to do so (Bosnia and Herzegovina). In Lebanon, the aim was to balance the power of the country's constituent groups within an integrated government. The post-*Ta'if* political consensus was to move on through silencing the past; discussions of blame, victimhood and responsibility were deemed dangerous for the fragile peace, an amnesty of war participants was declared and society entered a process of reconstruction which obscured the past through a collective amnesia; the political realm, formal education and institutions limited the impetus of civil society calls to discuss the war (Barak, 2007; Collings and Khalaf, 1994; Haugbolle, 2010; Larkin, 2012). The narrative of a 'war of others' on Lebanese territory obfuscated agency and responsibility of local actors, but also permitted a Lebanese state to continue its inherited *consociational arrangement* from the early years. Consequently, Beirut was reconstructed as the capital of a sole Lebanese state without a *de jure* territorial segregation between groups. In Bosnia and Herzegovina, in contrast, the internationally sanctioned peace agreement led to a loose federation of highly decentralized entities, which were *de facto* associated with particular ethnic groups. The Dayton Accords defined Bosnia and Herzegovina as a state composed of two entities: the Federation of Bosnia-Herzegovina and the Republika Srpska (RS). While the agreement did not stipulate territorial ethnic separation and it aimed to reverse the forced ethnic removals with a policy for displaced persons return, the two entities remained largely associated with post-war ethnic majorities and a political scene dominated by nationalist parties: a difficult Bosniak-Croat cohabitation in the Federation and Bosnian Serb overwhelming control over RS. Sarajevo, the capital of the state and, at the same time, of the Federation lay at the border between the two entities, with its southern quarters belonging to the RS, constituting a separate municipality, *Istočno Sarajevo* (Eastern Sarajevo). Reconstruction thus occurred in a seemingly undivided Beirut and an asymmetrically territorially divided Sarajevo.

The *Ta'if* accords in Lebanon and the charismatic figure of Prime Minister Rafiq Hariri translated into a strong, goal-oriented reconstruction process, aiming to recuperate Beirut city centre as a symbol of a strong, entrepreneurial Lebanese nation, united through its ambition to become again a leader of Middle Eastern economy. It reflected the geopolitical return of Lebanon in the Arab world and aimed to erase the urban geopolitics of war between military factions over urban territory. An *ad hoc* reconstruction company, the Lebanese Company for the

Development and Reconstruction of the Beirut Central District (Solidere), was established in 1994 as a joint-stock company in order to reshape Beirut's centre as an urban and political project. National Law 117 of 1991 was drafted especially for the creation of Solidere, attributing it prerogatives of planner, developer and manager of the city centre (Gavin and Maluf, 1996). Solidere was empowered to expropriate city centre property, compensating pre-war owners with company shares (interview A, Nabil Rached, 14 April 2009).[1] It drafted an integrated urban masterplan and demolished many remaining structures to (re)construct a high-standard designed city centre echoing the French mandate architecture, with immaculate public spaces and a homogeneous look (Gavin and Maluf, 1996; Humphreys, 2015; Ragab, 2011; Sawalha, 2010). The government-led reconstruction in the city centre was heavily criticized by civil society for its opacity, treatment of owners, destruction of still-standing heritage buildings in order to build a 'heritage' centre, as well as the neglect of other areas in the city (Ragab, 2011). The rest of the city witnessed a piecemeal reconstruction, with limited public funds and often-uncoordinated efforts; it was not part of the political-economic project of the brand of Beirut centre, and it included areas with strong political opposition to Hariri. In later decades, non-government actors such as Hezbollah took leadership in rebuilding Beirut's southern suburbs, particularly after the aerial attacks by Israel in 2006 (Fawaz, 2014; Harb, 2008, 2010; Saliba, 2013). Civil society, architect- and planner-led reconstruction plans for other neighbourhoods affected by the war, such as the *Elyssar* plan for southern Beirut, did not receive similar levels of support from the authorities, remaining incomplete or not implemented at all (Harb el-Kak, 2000).

While, for Beirut, the reconstruction of the city centre became a matter of national importance, there was no integrated plan of reconstruction for Sarajevo. The situation mirrored the piecemeal and privately led reconstruction of the Beirut periphery. Post-war Sarajevo has been administered by a complex administrative-territorial system; the 1986 masterplan remained valid, but the powers of the city mayor and planning were limited by the competencies of other territorial units, conceived as part of Dayton's democratization and decentralization agenda. Gordana Memišević, Head of the Research and Planning Department of the Sarajevo Kanton Planning Institute, described the paradox that while Sarajevo became more important as a capital of an independent state, its territory and its planning prerogatives reduced significantly (interview B, 14 July 2014). This is not only related to the division of the city (*grad*) between entities, but also to the emergence of intermediate planning levels, including the *kanton* (canton) – the intermediate administrative unit between entity and municipality (*opština/ općine*). Through a complicated process of matching international funders with sites that needed reconstruction, the *kanton* and the municipalities oversaw an urban reconstruction that was project-based and not integrated. The lack of an urban vision by most of the actors participating in the reconstruction was criticized by French architect Jean François Daoulas (interview C, 30 September 2009). For instance, reconstruction in the Dobrinja neighbourhood, in the southwestern part of Sarajevo, without considering the need for the adjacent Sarajevo airport to expand, condemned the airport to remain a small capacity facility (ibid.). The piecemeal, bandage approach is thus different from Hariri's vision of a Beirut that, through a gleaming new centre, would become a symbol of a new, confident Lebanon that could take back its place in the economic geopolitical system of the region. The drive of Hariri to remodel the centre of Beirut, with economic but also symbolic-geopolitical implications for the Middle East, did not find an equivalent to the same degree in post-war Sarajevo. Reconstruction in Sarajevo was not matched by the same geopolitical confidence: the international presence of the Office of the Higher Representative – at least in the first years – and the weak prerogatives of various state actors that Dayton enshrined (Bieber, 2006; Caplan, 2005; Knaus and Martin, 2003) did not permit such a vision or project to come to light.

Structural violence of reconstruction in contested cities

Shifting the scale from urban reconstruction to architectural reconstruction of particular buildings can also help in understanding the (geo)political underpinnings of reconstruction in contested cities. On the one hand, reconstruction can enhance geopolitical discourses of continuity. Robert Bevan (2007) discusses how what he terms *unintentional monuments* – places for everyday life, such as houses of worship, fountains, libraries – become, through their reconstruction, intentional monuments. They show that a community, or a particular group, have endured, survived, or triumphed over war. The reconstruction of the city centre of Beirut was such an intentional act of demonstrating that 'old' Beirut is back and can compete with this cultural capital and brand against 'new' Dubai. In Sarajevo, the reconstruction of mosques underlined the survival of the Bosniak majority, the reconstruction of common buildings maintained Sarajevo's urban imagination as a cosmopolitan hub.

Reconstruction of architectural objects can, on the other hand, symbolically prolong war through other means. Selective reconstruction and the construction of new structures that enhance the predominance of a certain group lead to a reshaping of the symbolic landscape. When only buildings related to the memory of one group are reconstructed and the others are left in ruins, this selective reconstruction can be translated geopolitically as a discourse of power. Similarly, when in the urban landscape one political group inserts an architectural object that is supposedly connected with its identity, this is another geopolitical act of dominance. Such objects of distinction and identity are, for Beirut and Sarajevo, their religious buildings, which become, in an urban geopolitical and symbolic reading, signifiers of groups and their presence in urban space. Beirut witnessed the painstaking reconstruction of religious buildings within the central district; nevertheless, Prime Minister Hariri supported and funded a sizeable Sunni mosque, a replica of Istanbul's Blue Mosque, which was built on the city's central Martyrs Square. Despite the claims of restoring an urban landscape of diversity, the large new mosque was seen by many Beirutis of a Christian or Shia Muslim background as a political statement from the Sunni politician. One of my interlocutors stated that the new mosque showed the Christians, whose quarter it faces, who is really ruling Beirut after the war (interview D, male, 46, Gemayzeh, 12 November 2009). This exemplifies how reconstruction can continue, through symbolic forms, the violence of war, recalling Dacia Viejo-Rose's (2011) argument of reconstruction as cultural violence, made with reference to Spain after the Civil War. Similarly, in Sarajevo, new religious buildings became space markers of who 'owns' the space, agents of group inclusion, but also symbols of a new exclusionary spatial hegemony. New mosques – and, to a lesser extent, Catholic churches in Federation Sarajevo, and Orthodox churches in East Sarajevo – not only respond to religious need, marginalized during socialism, but appropriate symbolic landscapes.

The insertion of capital from foreign countries brings in another geopolitical dimension. In Sarajevo, funders from Saudi Arabia, Indonesia and other Islamic countries supported grand architectural acts of mosques that dominate the landscape. Bosnian architect Isak Ćavalić commented that a "new line of badly behaved architects" kept introducing grand un-Bosnian elements in mosque architecture (interview E, 10 January 2014). Many Sarajevans, of diverse backgrounds, decry, in particular, the influence of Saudi capital in erecting grand new mosques and education centres, which also act as a place to propagate Salafi (*Wahhabi*) radical Islam perceptions. In their accounts, these mosques are 'foreign', as Bosnian Islam is 'different', echoing the geopolitical musings of Bosnia as a Western state with a Western type of Islam, distinctive from Arab conservative versions (Bougarel, 2007). Turkey's urban interventions are seen in a more sympathetic manner, though the post-colonial attitudes of Erdogan's government came under the scrutiny of some interlocutors. Architectural interventions were seen in larger geopolitical frameworks, as manifestations of allegiances and competitions, with discussions in the Bosnian media of how the government would favour Turkish, Saudi or US projects to display particular allegiances and invite further investment. Beyond investment comes the question of markets; in Beirut, interlocutors pointed out that the properties in the rebuilt city centre are mostly bought by Gulf Arabs, as a prized possession in the Lebanese capital, in the historic 'Paris of the Middle East'. Their owning of property and their rare appearance in the city also makes the centre a ghost town for most of the year.

Beyond architectural reconstruction, there is another instance of symbolic violence in post-war urban space reconfiguration, which comes in the form of memorials. With regards to the symbolic violence of memorialization, however, Beirut and Sarajevo differ greatly. Sarajevo abounds with memorials about the victims of 'Serbian criminals', while East Sarajevo has a number of memorials of Serb victims and heroes. The existence and ossification of national narratives about the war led to memorialization processes that are focused on the victimhood of one group and the blame of the other. In contrast, in Beirut, the political taboo of discussing the war led to the absence of memorials or even signs of war in the city centre. An interpretation of Solidere's clean slate is also a desire to erase any memory of the war. While in Sarajevo urban space is contested and appropriated through memorialization and reconstruction, with architecture and memorials embodying the discourse of conflict through symbolic violence, in Beirut reconstruction was shaped as a process of neutralization, privatization of space and amnesia-building, erasing the memory of the war and of contestation. It is a different from of symbolic violence of reconstruction: not one directed against another group, but one directed against the past itself.

Shaping places of inclusion and exclusion

Fregonese (2012) argued that geopolitical analysis should be more concerned with the everyday, the unofficial and the unplanned, as well with the urban geopolitics of peace and inclusion, rather than continuing to be dominated by discussions on war. Consequently, an analysis of everyday practices and of how peacetime inclusion and exclusion shape the city can advance our understanding of urban geopolitics. Divided Sarajevo consists now of two largely homogeneous separate urban units. In the Federation-controlled Sarajevo, aside from a small Croat minority, more than 80 per cent of population is Bosniak, including a significant number of war refugees. Istočno Sarajevo, in RS, is overwhelmingly Serb-populated and has been reconfigured with the purpose of creating an urban environment that provides an alternative to the 'lost' city centre of Sarajevo. Pedestrian areas throughout the city of Istočno Sarajevo aim to provide a replacement for the traditional walk – the korzo – on the main pedestrian drag in the Habsburg and Ottoman core of Sarajevo. Housing construction in Istočno Sarajevo has been high, while in the Federation many apartments lie unoccupied. Several interviewees in 2009 (F, G, H) accounted how they moved there to live among people of their own kind, as they worried about a perceived rise in radical Islam in Sarajevo proper. Yet the border between the two Sarajevos has become much more porous in recent years. Interviewees on the RS side (H, I) in 2009 and 2014 expressed how increasing numbers of Bosnian Serbs were seeking jobs in Federation Sarajevo. On the other side, a number of Bosniaks talked about how they travelled to Istočno Sarajevo to buy pork, which is rare to find in the city centre. Economic needs blur political boundaries in Sarajevo. Nevertheless, political decisions challenge the slow reintegration of the city through measures and regulations: in 2015, taxi companies registered on one side were denied the right to pick up or take passengers to the other Sarajevo, while there is also a lack of public transport in between the two.

In Beirut, while much of the residential segregation persisted after the end of war, there was no institutional separation between areas in relationship to their group majorities. Interviews in 2009 (J, K, L) featured discussions of a different social use of space depending on groups, but also of various practices that transcended these war-shaped individual boundaries. As for urban post-war reconstruction, the Solidere project led, at a first glance, to a reintegration of urban space divided by the Green Line during the war. In the reconstructed centre, a 'Garden of Forgiveness' emerged as a space of contemplation amid archaeological – not war – ruins, reminiscing all Lebanese about their common Phoenician

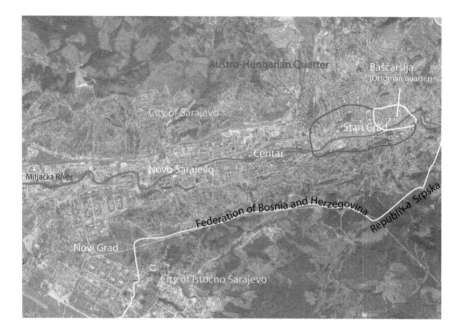

Figure 1.2 Sarajevo
Source: Author, 2015, based on Google Earth data

roots and the layers of history that came after. Flanked by two mosques and three churches of different Christian denominations, the Garden was imagined to define a pluralist Beirut (interview A, Nabil Rached, 14 April 2009). Severed from the rest of the city by iron fences guarded by armed soldiers, this picture-perfect downtown is, however, visibly an island. To enter this neatly landscaped urban realm one has to be scrutinized by soldiers and private security guards. Only people who look 'appropriate' are let in. There have been many instances when poorly dressed people or 'bothersome' types were not let in or were questioned by the guards (interviews L, N, local residents, April–November 2009). The promotional materials of Solidere proclaim the 'exclusive' nature of its developments as a brand (Solidere, 2009). Residential property is expensive and targeted at rich Gulf Arabs, who buy apartments for occasional use. The area consequently lost its pre-war role of a busy hub for all; in fact, many residents of Beirut often refer to the central area as 'Solidere', saying 'I don't usually go to Solidere' (interviews J, L, M, N, Beirut residents, April–November 2009).

At the margins of the reconstructed centre, the main square, Martyrs Square, has been lying unkempt, undesigned and underused for decades. However, it is this space that brought many together, with demonstrations of large proportions taking place here in the last decade, reflecting that the political role of public spaces has not been undermined by control, over-commercialization or neglect. Similarly, in 2014, Sarajevo's administrative centre has become the scene of

Figure 1.3 Beirut
Source: Author, 2015, based on Google Earth data

major demonstrations, challenging the heavily bureaucratic and inefficient state emerging out of the Dayton framework.

The large public political protests echo desires for more accountable political processes and highlight the lack of participation in decision-making. This is also reflective of reconstruction: in both cities, top-down decision-making trumped any consultation and participation; the agency of the local, both in terms of residents and many local urban planners and architects, was minimal. City-making was mired in national-level politics and state geopolitical undertones; while war brought in the urban geopolitics of fighting for territory or for power, the cities were remade in relationship to their capital status, embodying national narratives and geopolitical discourse – Beirut centre as a window of opportunity in the Middle East, while its peripheries fragmented, mirroring local politics; Sarajevo as a majority Bosniak capital of a contested state and a periphery belonging to a straying entity, with architectural interventions to demarcate territorial stakes and claims.

Conclusion

Although similar in their pre-war and war situations, Beirut and Sarajevo started to evolve differently following the different political consensus that dominated post-war Lebanon and Bosnia and Herzegovina. While Bosnia and Herzegovina followed the Dayton Accords and institutionalized separation, Lebanon aimed after the *Ta'if* Agreement to balance the power of its constituent groups within an integrated government. State politics had a great impact on the cities themselves, including the manner in which architectural reconstruction took place. On the one hand, in Beirut, the political will of Prime Minister Hariri aimed to create a new 'old' downtown with a stated goal to put Beirut and Lebanon again at the forefront of the Middle East, as well as to unite people through the rebuilding of common places such as the Beirut souqs or public spaces projects such as the Garden of Forgiveness. Sarajevo, on the other hand, evolved into two separate cities (the Federation-controlled Sarajevo proper and Eastern Sarajevo, in RS), where architectural reconstruction functioned as a piecemeal repair, while promoting particular religious buildings and memorials. International capital, particularly from Saudi Arabia and other Muslim countries, had a direct impact in the shaping of places in the two cities, favouring certain expressions of identity over others. The different political frameworks of reconstruction thus provided for distinct urban outcomes.

The reconstructed urban environment in both cities, Beirut and Sarajevo, also contributed to the shaping of territorialization and both national and urban geopolitical imaginations. Religious buildings marked the place appropriation of groups over territory. Memorials in Sarajevo brought forward discourses of victimhood and blame. In Beirut, what can be seen as a reconstruction favouring the integration of groups and the promotion of a unified city, but a unified 'nation', confident in its economic outlook, has created other types of divisions. Religious divisions gave way, in Beirut, to class barriers – in separating a city centre for the rich and the foreign and a periphery for the poor and the local. Similarly, in Sarajevo, a political structure favouring separate lives is slowly challenged by a reintegration expressing economic needs and realities.

While Sarajevo and Beirut shared a number of characteristics – nature of ethno-religious and linguistic mix; urban imaginaries of cosmopolitan pluralism in contested states with antagonistic narratives; historical influences; a reflection of divided city/contested city aspects – the political arrangements after war in their respective countries led to different urban outcomes. The cities' transformation, in this case, had less to do with local, urban agency than with that of states. It does not imply that contested cities mirror the country-scale processes, but it shows that the impact of state arrangements is still very important for contested cities and their local urban geopolitics. While the urban imaginaries of the two cities for their populations were centred on cosmopolitanism and coexistence, a frozen antagonistic coexistence on one side and a division, as opposed to indivisibility and antagonistic blaming in the other differentiated the urban outcomes. As such, post-war reconstruction in these contested cities prolonged and transformed the nature of contestation of urban space in relationship to larger political frameworks and geopolitical discourse. Consequently, while agreeing that a focus on the urban condition is indispensable if one wants to understand ethno-national and religious conflict (Pullan and Baillie, 2013), the reflection on Beirut's and Sarajevo's reconstruction underlines the importance of analysing different scales. The reconfiguration of urban environment is mired in different geopolitical imaginations of the state, international actors (i.e. neighbours) or the international community at large, which determined post-war political arrangements or influenced reconstruction through capital flows. This chapter highlights the need for more multi-scalar layering of urban geopolitics and the analysis of urban processes and outcomes in relationship to different dimensions and scales of contestation.

Note

1 List of interviews: A – Nabil Rached, Solidere, Beirut, 14 April 2009; B – Gordana Memišević, Head of the Research and Planning Department of the Sarajevo Kanton Planning Institute, Sarajevo, 14 July 2014; C – Jean François Daoulas, architect, Sarajevo, 30 September 2009; D – anonymous, male, 46, Gemayzeh, Beirut, 12 November 2009; E – Isak Ćavalić, architect, Sarajevo, 10 January 2014; F – anonymous, male, 45, pension owner, Pale, East Sarajevo, 22 September 2009; G – anonymous, female, 40, East Sarajevo, 23 September 2009; H – anonymous, male, 62, East Sarajevo, 15 January 2009; I – anonymous, female, 65, East Sarajevo, 22 July 2013.; J – anonymous, male, Hamra, 28, Beirut, 13 November, 14 November 2009; K – anonymous, female, 26, Achrafiyeh, Beirut, 17 November 2009; L – anonymous, male, 35, Bourj Hammoud, Beirut, 25 October 2009; M – anonymous, female, 26, Jounieh, Beirut, 10 November 2009; N – anonymous, male, 28, South Beirut, 16 April 2009.

References

Bakshi, A. (2011) Memory and place in divided Nicosia. *Spectrum: Journal of Global Studies*, 3(4): 27–40.

Bakshi, A. (2014) Urban form and memory discourses: spatial practices in contested cities. *Journal of Urban Design*, 19(2): 189–210.

Barak, O. (2007) 'Don't mention the war?' The politics of remembrance and forgetfulness in postwar Lebanon. *The Middle East Journal*, 61(1): 49–70.

Barakat, S. (ed.) (2005) *After the Conflict: Reconstruction and Development in the Aftermath of War*. London: I. B. Tauris.

Barakat, S., Car, Z. A., and Halls, P. J. (2008) GIS methodologies in post-war reconstruction. In S. Wise and M. Craglia (eds), *GIS and Evidence-Based Policy Making*, Boca Raton, FL: CRC Press: 261–82.

Bevan, R. (2007) *The Destruction of Memory: Architecture at War*. London: Reaktion.

Bieber, F. (2006) After Dayton, Dayton? The evolution of an unpopular peace. *Ethnopolitics*, 5(1): 15–31.

Bogdanović, B. (1993) *Die Stadt und der Tod: Essays*. Klagenfurt: Wieser.

Bollens, S. A. (2006) Urban planning and peace building. *Progress in Planning*, 66(2): 67–139.

Bollens, S. A. (2012) *City and Soul in Divided Societies*. London and New York: Routledge.

Bougarel, X. (2007) Bosnian Islam as European Islam: limits and shifts of a concept. In A. Al-Azmeh and E. Fokas (eds), *Islam in Europe: Diversity, Identity and Influence*. Cambridge: Cambridge University Press.

Brand, R., and Fregonese, S. (2013) *The Radicals' City Urban Environment, Polarisation, Cohesion*. Farnham: Ashgate.

Calame, J., and Charlesworth, E. (2009) *Divided Cities: Belfast, Beirut, Jerusalem, Mostar, and Nicosia*. Philadelphia: University of Pennsylvania Press.

Caplan, R. (2005) Who guards the guardians? International accountability in Bosnia. *International Peacekeeping*, 12(3): 463–76.

Charlesworth, E. R. (2006) *Architects Without Frontiers: War, Reconstruction and Design Responsibility*. Amsterdam and London: Architectural.

Collings, D., and Khalaf, S. (eds) (1994) Culture, collective memory, and the restoration of civility. In D. Collings (ed.), *Peace for Lebanon? From War to Reconstruction*. Boulder and London: Lynne Rienner.

Coward, M. (2006) Against anthropocentrism: the destruction of the built environment as a distinct form of political violence. *Review of International Studies*, 32(3): 419–37.

Coward, M. (2009) *Urbicide: The Politics of Urban Destruction*. London: Routledge.

Donia, R. J. (1994) *Bosnia and Hercegovina: A Tradition Betrayed*. London: Hurst.

Donia, R. J. (2006) *Sarajevo: A Biography*. Ann Arbor: University of Michigan Press.

Driessen, H. (2005) Mediterranean port cities: cosmopolitanism reconsidered. *History and Anthropology*, 16(1): 129–41.

Fawaz, M. (2014) The politics of property in planning: Hezbollah's reconstruction of Haret Hreik (Beirut, Lebanon) as case study. *International Journal of Urban and Regional Research*, 38(3): 922–34.

Fregonese, S. (2009) The urbicide of Beirut? Geopolitics and the built environment in the Lebanese civil war (1975–1976). *Political Geography*, 28(5): 309–18.

Fregonese, S. (2012) Urban geopolitics 8 years on: hybrid sovereignties, the everyday, and geographies of peace. *Geography Compass*, 6(5): 290–303.

Freitag, U., and Lafi, N. (2014) *Urban Governance Under the Ottomans: Between Cosmopolitanism and Conflict*. Abingdon: Routledge.

Freitag, U., Fuccaro, N., Lafi, N., and Ghrawi, C. (2015) *Urban Violence in the Middle East: Changing Cityscapes in the Transformation from Empire to Nation State*. New York: Berghahn.

Gavin, A., and Maluf, R. B. (1996) *Beirut Reborn: The Restoration and Development of the Central District*. London and Lanham, MD: Academy Editions, distributed by National Books Network.

Graham, S. (2004) *Cities, War and Terrorism: Towards an Urban Geopolitics*. Oxford: Blackwell.

Harb, M. (2008) Faith-based organizations as effective development partners? Hezbollah and post-war reconstruction in Lebanon. In G. Clarke and M. Jennings (eds), *Development, Civil Society and Faith-Based Organizations: Bridging the Sacred and the Secular*. Basingstoke: Palgrave Macmillan: 214–39.

Harb, M. (2010) *Le Hezbollah à Beyrouth, 1985–2005: de la banlieue à la ville*. Paris and Beirut: Karthala, Institut français du proche-orient.

Harb el-Kak, M. (2000) Post-war Beirut: resources, negotiations, and contestations in the Elyssar Project. *The Arab World Geographer*, 3(4): 272–88.

Haugbolle, S. (2010) *War and Memory in Lebanon*. Cambridge: Cambridge University Press.

Hayden, R. M. (2013) Intersecting religioscapes and antagonistic tolerance: trajectories of competition and sharing of religious spaces in the Balkans. *Space and Polity*, 17(3): 320–34.

Hudson, M. C. (1997) Trying again: power-sharing in post-civil war Lebanon. *International Negotiation*, 2(1): 103–22.

Humphreys, D. (2015) The reconstruction of the Beirut central district: an urban geography of war and peace. *Spaces and Flows: An International Journal of Urban and ExtraUrban Studies*, 6(4): 1–14.

Ilbert, R. (1996) *Alexandrie, 1830–1930: histoire d'une communauté citadine*. Cairo: Institut français d'archéologie orientale.

Jamaković, S., and Association of Architects (1993) *Warchitecture*. Sarajevo: ARH.

Knaus, G., and Martin, F. (2003) Travails of the European Raj. *Journal of Democracy*, 14(3), 60–74.

Larkin, C. (2012) *Memory and Conflict in Lebanon: Remembering and Forgetting the Past*. London: Routledge.

Lyons, M., Schilderman, T., and Boano, C. (eds) (2010) *Building Back Better: Delivering People-Centred Housing Reconstruction at Scale*. Rugby: Practical Action.

Mazower, M. (2005) *Salonica, City of Ghosts: Christians, Muslims, and Jews, 1430–1950*. New York: Alfred A. Knopf.

Naimark, N. M., and Case, H. (2003) *Yugoslavia and its Historians: Understanding the Balkan Wars of the 1990s*. Stanford, CA: Stanford University Press.

Nasr, D., Massoud, M. A., Khoury, R., and Kabakian, V. (2009) Environmental impacts of reconstruction activities: a case of Lebanon. *International Journal of Environmental Research*, 3(2): 301–8.

Potyrala, A. (2009) Sarajevo: from the ashes of conflict to cold coexistence. In J. Janczak (ed.), *Conflict and Cooperation in Divided Cities*. Berlin: Logos.

Pullan, W. (2013) Spatial discontinuities: conflict infrastructures in contested cities. In W. Pullan and B. Baillie (eds), *Locating Urban Conflicts: Ethnicity, Nationalism and the Everyday*. Basingstoke: Palgrave Macmillan.

Pullan, W., and Baillie, B. (2013) *Locating Urban Conflicts: Ethnicity, Nationalism and the Everyday*. Basingstoke: Palgrave Macmillan.

Pullan, W., Sternberg, M., Dumper, M., Larkin, C., and Kyriacou, L. (2013) *The Struggle for Jerusalem's Holy Places*. Abingdon: Routledge.

Ragab, T. S. (2011) The crisis of cultural identity in rehabilitating historic Beirut-downtown. *Cities*, 28(1): 107–14.

Ragette, F. (1983) *Beirut of Tomorrow: Planning for Reconstruction*. Beirut: American University of Beirut.

Ramet, S. P. (2006) *The Three Yugoslavias: State-Building and Legitimation, 1918–2005*. Washington, DC, and Bloomington: Woodrow Wilson Center Press; Indiana University Press.

Salamey, I. (2013) *The Government and Politics of Lebanon*. London: Routledge.

Saliba, R. (1998) *Beirut 1920–1940: Domestic Architecture Between Tradition and Modernity*. Beirut: Order of Engineers and Architects.

Saliba, R. (2013) Studio culture/war culture: pedagogical strategies for reconstructing Beirut's southern suburbs. *Architectural Research Quarterly*, 17(2): 167–76.

Salibi, K. S. (1988) *A House of Many Mansions: The History of Lebanon Reconsidered*. London: I. B. Tauris.

Sawalha, A. (2010) *Reconstructing Beirut Memory and Space in a Postwar Arab City*. Austin: University of Texas Press.

Solidere (2009) *The Quarterly*. Available from: http://www.solidere.com/corporate/publications/quarterlies. Accessed 21 April 2016.

Viejo-Rose, D. (2011) *Reconstructing Spain: Cultural Heritage and Memory After Civil War*. Brighton and Portland, OR: Sussex Academic Press.

Woods, L. (1997). *Radical Reconstruction*. New York: Princeton Architectural Press.

2 Negotiating cities

Nairobi and Cape Town

Liza Rose Cirolia

Introduction: between extremes

In recent years, there has been substantial effort within and beyond academia to make sense of the development of African cities. The dominant literature on African cities is highly polarized. Journalists often present African cities as sites of despair and chaos, chronicles of oppression and poster children of planning 'gone wrong' (Pieterse, 2008; Murray, 2008). In contrast, international consultancy firms and select development agencies prefer the 'Africa rising' narrative, celebrating the rising GDP, middle class and infrastructure investment evident in many African cities (e.g. McKinsey and Company, 2010, 2012).

Critiquing this narrow conceptualization of 'southern cities', Lemanski and Oldfield (2009: 634) call for work which explores cities "as complex and contradictory sites in which diverse residents and urban processes function in the context of state (dis)engagement". Heeding this call and steering a course between the afro-apocalyptic and afro-euphoric narratives which plague the debates on African cities (ibid.), this chapter identifies and describes the planning frameworks and patterns of urban real estate development at the city scale in two cities, Cape Town, South Africa, and Nairobi, Kenya. It shows that, in contrast to narratives of chaos and despair, there are clear patterns shaping the development of both cities. However, none of these development trends can be easily classified as 'ideal development typologies' in line with internationally recognized planning principles or even local plans.

In both cases, these emerging patterns represent tense negotiations between various implicated development actors, challenging planning and infrastructure conventions and local policies. Spaces of negotiation and respective power differentials differ in the two cities, but both cases show that city planning is an iterative and political effort, the outcomes of which are determined not by resolved urban ideals, but by ongoing compromise and conflict. Putting into conversation the experiences of these cities can offer substantial insights, not only on the diversity of development experience evident in African cities, but also on ways in which the state, land development and infrastructure coproduce a negotiated urban geopolitical fabric in a context where traditional municipal finance and land development principles and conventions fail to apply.

This chapter is based on a desktop analysis of plans and policies, as well as interviews undertaken in each city. Interviews were conducted with city planners (public servants and private sector), housing officials, developers, local NGOs and infrastructure specialists. Between 20 and 30 interviews were conducted in each city between 2013 and early 2015.[1] The following section frames the study, focusing on the value of comparing cities and the theme of urban development. The body of this chapter unpacks the key question 'What drives the production of urban fabric?' In response to this question, three development patterns are identified for each city. The chapter concludes with a discussion on the value of these cases for understanding the contested nature of urban (re)production.

Thinking about cities and their (re)production

While northern academia remains the epicentre of urban knowledge production, a growing body of work points to the need to challenge this hegemony. Simple narratives on developing cities, generally, and African cities specifically, are increasingly rejected by authors who call for more detailed and less rhetorical work (Pieterse, 2008; Gilbert, 2009; Myers, 2014). Academics call for work which is empirically grounded and seeks to understand cities 'on their own terms'. This is particularly imperative for cities that have historically not been sites of comparison, theorization and knowledge building – what are commonly called 'southern cities', though much debate exists as to the usefulness and definition of this term (Myers, 2014; Robinson, 2016; Oldfield, 2014).

The epistemic implications are that researchers concerned with the production of knowledge not only on, but also *from* the South (i.e. peripheral sites of knowledge production) often begin their research by exploring that which is actually happening in cities. There is a growing body of literature that works to produce understanding and knowledge through inductive frameworks. This gesture is not (one hopes) the search for endless specificity and description, but rather the desire to articulate generalities previously overlooked. It is within this frame that the study of urban development is discussed.

Undeniably, cities require the ongoing production (or regeneration) of space (Healey and Barrett, 1990). Part of this production involves the material development of infrastructure and urban property. Within conventional property studies, planning discourses and even economic rationale, the roles and responsibilities of players are clear and the development process is linear. The property developer (be they an individual, firm, or community) is responsible for financing the physical buildings and internal services on their property. The state, in contrast, is responsible for the services external to the development site, bulk and connector infrastructures which link properties through complex material pathways and flows (Callies and Grant, 1991; Kihato, 2012; Estache, 2010). The planning sector is responsible for determining if particular developments can be sustained by the capacity of the bulk systems, as well as if the development is 'desirable' for the long-term public good of the city.

The logic underpinning this 'ideal' urban development scenario is that the state invests in planning framework and in infrastructures. This predictability and provision increases the value of land and property by lowering the risk to financiers and developers and creating developable land. This builds value that can be captured over time by the state (in the form of property rates and other taxes and levies) and can be reinvested in infrastructure and operations, thus creating a virtuous system of urban development (Ingram *et al.*, 2013; Franzsen, 2003).

Unsurprisingly, this narrative renders most developing cities 'failures', unable to sustain this trajectory and unwilling to follow the necessary prescriptions for success. Undeniably, many African cities have not achieved this so-called virtuous cycle. This requires, therefore, a fundamental rethinking of the role of the state in urban development. In many large African cities there has been no central entity effectively controlling development or upholding a 'public mandate' to invest in infrastructure (Kihato, 2012; Parnell and Pieterse, 2014). From informal water vendors to solar panels, mini-bus taxis (*matatus*) to boreholes, the responsibility of public service provision increasingly falls outside the state's ambit, disabling any legitimate claim to this value creation on the part of the state. Many authors celebrate the ingenious ways in which spatial decision-making, service delivery and infrastructure access is achieved in the absence of city-scale provision, while simultaneously recognizing the limitations and challenges these practices present (Amin, 2014; Simone, 2010; Roy, 2005). Making sense of this tension requires a more complex understanding of legitimation of city-making processes evident in southern cities.

These experiences help us to understand that the physical development of the city, and particularly the way in which the state interfaces with development and developers, is not merely to understand the technical chores of planners, architects, engineers and developers (Napier *et al.*, 2013; Ennis, 1997). Instead, it is to unpack the many sites of power, politics and capital circulation as they embed in and produce space – in short, the geopolitical processes of urban development (Graham, 2010; Harvey, 2012; Murray, 2008; Avni and Yiftachel, 2014). It is to recognize a process born out of complex social, economic and political negotiation, where those who supply, demand, broker and regulate space, coproduce the material and spatial outcomes of the city (Alonso, 1960; Guy and Hanneberry, 2002; Bekker and Therboro, 2011). Echoing these political and spatial gestures, the central role that powerful interest groups (and at times the state) play in constraining (or not) land, labour, capital, development rights and technologies, ultimately shaping the outcomes of property development, should be of central concern (Healey and Barrett, 1990; Graham, 2010; Adams and Tiesdell, 2010). Within these discussions, land and property are not merely the stage on which urban processes are played out; the material development of these spaces themselves constitutes and forms part of the production and consumption processes of cities (Healey and Barrett, 1990).

In order to look more closely at the grounded practices of urban geopolitical (re)production, processes underway in Nairobi and Cape Town are unpacked below.

Nairobi: lifestyle estate, tenements and land-buying companies

Nairobi was established in 1899 as a rail depot along the Uganda Railway, or the 'Lunatic Line', as the British press regularly called it (Martin, 2006). Today, Nairobi is the capital of Kenya and the largest city in the region, with over 5 million inhabitants in the metropolitan area and an annual growth rate of over 4 per cent.

Since independence, Nairobi has seen a succession of urban plans which have aimed to guide the development of the city. These include the Nairobi Metropolitan Growth Strategy of 1973, the Nairobi City Development Ordinances and Zones, the Spatial Planning Concept for Nairobi Metropolitan Region and the Nairobi Metro 2030 policy (Owuor and Mbatia, 2011; Huchzermeyer, 2011). Under the current constitutional dispensation, the functions of forward planning and development control have been devolved to county governments ([section 104 (1)] of the County Government Act of 2012). Within the newly formed Nairobi City County (NCC), the Department of Lands, Physical Planning and Housing is charged with these functions. This means that the local government has become responsible for ensuring that the development of the city, particularly in terms of infrastructure and real estate, aligns with the city plans. The city is currently drafting the Integrated Urban Development Master Plan (NIUPLAN), to replace the 1973 strategy. The draft NIUPLAN is an impressive document which draws together substantial data on the city of Nairobi (Nairobi City County, 2014). However, the plan makes no reference to what city officials and local developers argue are the dominant real estate development processes underpinning the growth of the city.

This section unpacks three important trends emerging in Nairobi. These trends work together to tell a story about the development of the city. First, tracks of land are being developed by large-scale private real estate companies, often backed by global financiers. Second, the infill of the city is largely being built by tenement landlords constructing multistorey rental accommodation for the lower and middle classes. Finally, the outskirts of Nairobi are being developed by land-buying companies (LBCs), established to purchase (and ostensibly develop) peri-urban land for its shareholders. While there are many variations on these processes, these high-level trends have profound implications for the shape and fabric of the city.

Private real estate companies building for the wealthy

Like many African cities, the demand for high-income property development in Nairobi has grown dramatically (Grant, 2015; Lemanski and Oldfield, 2009). Many high-end developers see Nairobi as the entry point to the continent from the East and an exciting investment destination (Hass Property Index, 2009). There are many opportunities to be critical of Nairobi's high-end and luxury development (Myers, 2015). However, the scale of planning and construction for the middle and upper class cannot be ignored or mocked for its extravagant architecture or aspirations. Nor can the investments which governments make to support these developments be written off as simply articulations of neoliberal globalization (ibid.) or world-class city fantasies (Watson, 2014).

While high-rise luxury developments in and around the core of the city have been a feature of Nairobi's development for decades, the more recent trend towards constructing middle-income 'lifestyle communities' and 'new cities' – which are actually suburbs – on greenfield sites cannot be ignored (Kihoro, 2015). These projects tend to be strategically located along the newly built super highways and bypass roads in and around Nairobi. In these projects, developers service and subdivide the land into residential and recreational plots in accordance with their site master plans. Houses are sold 'off the plan' in phases, increasing in cost as the project progresses. In most of these projects, on-site infrastructure provision, such as boreholes and solar energy, fill the gap in municipal services. Thika Greens Golf City, for example, boasts a 206-metre borehole which is pumped using solar energy.

The 'lifestyle' benefits and development opportunities of buying in these communities are commonly portrayed through promotional videos and intricate site master plans. The aforementioned Thika Greens Golf City, for example, plans to have a conference facility, retirement village and a range of special golfing features. Similarly, Garden City, developed with investment from the International Finance Corporation (IFC) and the Commonwealth Development Corporation (CDC), plans to construct the largest mall in East Africa and the first 'integrated residential development' in Kenya. The Four Ways Junction, which is explicitly modelled after South Africa's 'security estates' (security-heavy gated communities), includes a country club and hotel. The Two Rivers Development Project, in line with the trend on the continent, includes a mall, a waterfront, residential apartments, two hotels and a hospital, largely targeted at Kenya's upper classes and international investors.[2]

Tatu City is perhaps the most dramatic expression of this development pattern (Watson, 2014; Grant, 2015). Tatu City, on 2,500 acres of land located on an old coffee farm in Kiambu County, was once a peripheral farm over 30km from Nairobi. However, the new Thika Superhighway has opened up possibilities for the development. The Tatu developers have been negotiating with Kiambu local government to fund a complete overhaul of the water and sanitation system. While the first phase of development was scheduled for 2013, Tatu City has experienced long delays in the uptake of the development due to protracted court cases over land rights and investment arrangements (between the participating coffee farm owners and the international investors, Rendevour). By the end of 2015, the site remained empty, save some promotional pillars.

Because the state offers little in terms of services, coupled with the espoused necessity of attracting global capital, it is difficult for the local government to effectively regulate developers of high-end mega-projects. In the case of Tatu City, the developers plan to take over local service provision, a manoeuvre only possible because they are administratively outside Nairobi. In Nairobi proper, in contrast, the only instrument which the local government commonly uses to push back against developers is the implementation of what the County planning department refers to as 'Planning Gains' (DFID, 2015). When the County approves building plans, *ad hoc* conditions are placed on the developer. As part of these conditions, the developer might be required to upgrade trunk or connector infrastructure surrounding the site (such as widen the roads) or to construct social infrastructure like schools or parks.

Tenement landlords

Much attention has been given to small-scale landlordism and slum development in Nairobi (e.g. Huchzermeyer, 2008; Amis, 1984; Gulyani and Talukdar, 2008; Gatabaki-Kamau and Karirah-Gitau, 2004; Parsons, 2013). Informal settlements, like many cities in the Global South, remain an important housing option for the poor. However, their expansion has been limited by land constraints and planning regulations. Instead, the ever rising importance of *tenement housing* for poor and lower middle-class residents represents an important shift in low-income property development (Huchzermeyer, 2011; Mwau, 2013).

Despite the obvious scale of this phenomena to anyone driving around Nairobi, there is little data on tenements. In 2009 the local government estimated that there

Figure 2.1 Tatu City
Source: Author, 2015

were over 10,000 buildings, mainly existing in Eastlands and parts of the Westlands area, with more being constructed daily. Huchzermeyer (2011) refers to Nairobi as a 'tenement city' due to the pervasiveness and diversity of tenement housing in the modern urban fabric. In some areas, these tenements form part of the diversity of building typologies, located on road reserves, setbacks, or public spaces within existing suburb; however, often the pressure to densify has created entire districts marked by high-rises which fail to conform to building and planning codes (Mwau, 2013). However, not all of the tenement developments are poor quality and hazardous. Areas such as Umoja Inner Core and Zimmerman are examples of high-quality semi-formal construction and an attempt to create attractive façades aimed at attracting middle-class residents (ibid.).

Recognizing its importance, the local government has historically turned a blind eye to the tenement construction in the city, neither planning for it nor moving to demolish it at scale (this development is completely ignored in the city plans). However, the frequent collapse of these buildings (resulting in a number of deaths) has compelled the local government to take some action. According to officials, Nairobi's planning department has deployed a technical team to rove the city and assess structures retroactively for their compliance with building standards (DFID, 2015). The intention is to minimize demolition and enforce basic engineering standards, rather than move to comply with plans or planning

Figure 2.2 A tenement development in Nairobi
Source: Author, 2015

standards, which would clearly be an impossibility. While the impacts of this programme are constrained by the bribery of inspectors and the limited capacity of the inspection unit, the programme represents an important negotiation between small- and medium-scale property developers and the state.

Collective buying and development

In Nairobi, a common way for land development to take place is through LBCs. Many authors have noted the importance of these companies in assisting ordinary families, operating in groups, to gain access to land (Yahya, 2002; Huchzermeyer, 2011). The LBC's approach involves a group of people pooling their savings to purchase a piece of land. The members of the cooperative are given 'share certificates' with the expectation of future subdivision and formal titling to follow. According to an interview with a representative from National Cooperative Housing Union (NACHU), in order to get the most value for money, this land tends to be without services and on the outskirts of the city where land costs are lower (DFID, 2015). This land-buying practice has been slowly expanding the urban fabric into the rural surrounds, representing a radical, if incremental, force in the spatial development of the edge of the city.

Once land is purchased, it is common for development of the site to take place quickly, despite the lack of infrastructure and rural zoning. Individual households rapidly erect structures and utilize individual infrastructure solutions. While established to service and develop the land, after settlement the LBC often becomes defunct (Taylor, 2004). Non-delivery of shared infrastructure has become the norm, rather than the exception, as companies increasingly establish with the mal-intent of serving the interests of elites or in neglectful haste to get low-income households onto the property ladder (Rakodi, 2007; Medard, 2010). In these cases, individual households build septic tanks and dig boreholes to provide basic services, essentially living off the grid.

As per the legislation, if infrastructure is not provided, the settlement stays in limbo, unable to be formally subdivided (as infrastructure is a prerequisite for these planning processes), granted occupation licensees, or be incorporated into the rates system. However, according to city planners, in many cases the city government has retroactively delivered bulk and even internal infrastructure to these peripheral settlements, largely with the incentive to bring large parts of the city into the tax net, but also as a means by which to leverage political support. In other cases, planners have agreed to incorporate these settlements into the urban fabric without infrastructural services, with the hope that formalization will allow for households to access formal credit sufficient to invest in the needed services.

Reflecting on Nairobi

On the surface, Nairobi's urban development appears to be a story of extremes and an exaggerated form of what Graham and Marvin (2001) call 'splintering urbanism'. However, upon deeper inspection there is much more nuance.

Nairobi's urban development story highlights the multi-dimensional nature of informality in the city. While Nairobi's slums remain important, the creeping informality of lower- and middle-income households on the edges of the city and tenement developments within its established neighbourhoods begin to reflect this nuance. Similarly, the story of aspiration – from golf estates to Tatu City – capture a more textured picture of the rising African middle-class resident and consumer. The state has not been fully in control of the spatial development of the city. However, arms and branches of the government have been acutely aware of the processes at play, working to negotiate urban outcomes with LBCs, tenement owners and elite developers alike.

Cape Town: 'RDP' housing, inner city regeneration and gated estates

South African cities are, from a planning perspective, best known for building racial segregation into the fabric of urban areas (Mabin and Smit, 1997; Harrison *et al.*, 2008). In Cape Town, apartheid planning included the creation of zones for white, coloured and black African households, forcibly removing those families and communities which did not align with the plan. Throughout the apartheid period (from the 1950s to the mid-1990s), the state built scaled developments, following in the footsteps of the British new towns approach, in peripheral townships in an effort to prop up a labour force necessary for sustaining urban areas. Statutory 'Guide Plans', blue-print land use plans drawn by the central government, served as the basis for planning decisions (Watson, 2003).

Efforts to create new urban plans and planning framework, which could undo the apartheid legacy, formed the basis of new dispensation. Most notably, the five-year Integrated Development Plan (IDP) tool was introduced in 1996 (Harrison, 2006). The IDP, despite many critiques, has become a staple in Cape Town's urban planning processes (Smit, 2015). Following suit, Cape Town created the Municipal Spatial Development Framework (1998) which reflected the ideal of township regeneration through a 'nodes and corridor' framework. Without statutory power or political will, however, the plan was largely ignored. In 2012, Cape Town created the City Development Strategy and the Spatial Development Framework (SDF). Despite these planning efforts, Cape Town remains divided along racial lines, value and opportunity concentrating in historically white areas. There are three important trends in the urban development of the city. First, the large-scale state-driven housing programme; second, inner city regeneration; and, finally, the rise of gated communities and suburbs.

State-driven residential expansion

A unique feature of South African urban expansion has been the rapid production of fully subsidized housing (colloquially referred to as RDP housing after the now defunct Reconstruction and Development Programme). As part of the RDP programme, in 1994, South Africa embarked on an ambitious drive to deliver free housing to the urban poor through the relocation of those living in shacks and overcrowded conditions to new housing projects. This housing predominantly takes the form of 'greenfield' low-density developments on the periphery of towns and cities where land is the cheapest and easiest to acquire. Over the last 22 years, the national government has built over 3 million units in South African urban areas, in contrast to the private sector which have developed just over a million new dwellings.

Cape Town's planning department have been at the forefront of critiquing RDP housing – in particular, for its contribution to urban sprawl. The Spatial Planning and Urban Design Department (SPUD) engages in combat with the housing departments in the City and Province, arguing for the upgrading of existing informal areas (through planning incorporation and incremental service delivery), and the support for rental accommodation (both formal and informal). However, the city's drive to meet delivery targets (set by the national government) tends to undermine these efforts.

While the housing programme has, undoubtably, created flat and relatively homogeneous suburbs, these areas have experienced incremental densification. Like Nairobi, land occupations by the urban poor were important forces in determining the structure of the city (Thorn and Oldfield, 2011). However, more recently, instead of settling new developments, the majority of expansion has taken place in the backyards of RDP houses. Despite the perpetual resistance of the local authorities (tasked with enforcing building and planning standards), households in possession of RDP housing almost always construct two to four additional dwellings on their plots.

Figure 2.3 Incremental extensions to RDP dwellings
Source: Warren Smit, 2015

The servicing of backyard properties as a constitutional right has been a major issue of debate, with Cape Town leading in the provision of permanent and temporary services for backyard dwellers. As a recent manoeuvre, permanent services are being provided to backyarders on public land (council-owned properties) and temporary services (portable toilets and shared stand pipes) are being provided to backyarders living in RDP projects. This servicing represents state efforts to come to grips with informal densification practices, the challenge of investing state infrastructure on private property and the role of the 'developmental local government' in the upholding of rights to access to essential services. If, as Pieterse and Cirolia (2016) argue, the RDP housing programme has been the *de facto* planning policy of the city – driving the shape and form of development due to its incredible scale and political power – the backyarding which accompanies it has turned urban citizens into small-scale developers with whom the state must reckon.

Inner city regeneration: knock on effects

Over the past 30 years, the City of Cape Town has attempted to improve the Central City, known as the central business district, or CBD (Visser and Kotze, 2008). After a period of decline in the late 1990s and the rapid loss of property rates revenues and businesses from the Central City (some of which moved to new development areas like Century City), the city government, fearing that Cape Town's inner city would follow in the footsteps of Johannesburg's 'white flight', established the Cape Town Partnership, a public–private partnership agency tasked with addressing the 'crime and grime' and disinvestment in the city. The Cape Town Partnership established the City Improvement District in 2000, drawing extra levies for the funding of security and waste collection (Didier *et al.*, 2012). The implementation of an area-based improvement programme can be seen, in part, as a negotiation or backlash against the 'one city one tax base' movement which led to the creation of the unicity in 2001 and which had, in essence, worked to redistribute revenue across the metropolitan area (ibid.).

The knock-on effect of this round of improvement efforts and investment has had a substantial impact on the value of properties in the city and surrounding residential suburbs, most notably in Sea Point and Woodstock – adjacent areas with old building stock which have historically been mixed-income areas (Pirie, 2007). These areas have experienced rapid redevelopment, with high-end urban developers (in contrast to their more suburban counterparts) buying older residential and industrial buildings and converting them to trendy upmarket apartments.

Along historic corridors, areas which have typically housed established low-income communities, immigrant communities and first-time home buyers, the CBD improvement has created a shift in demographics and architectural aesthetic, no longer affordable to these market segments. Concern with this process has primarily been documented by critical scholars and activists interested in the concept of gentrification and how it is played out in South African cities (Visser and Kotze, 2008; Miraftab, 2005; Bremner, 2000; Wilson, 2011). However, the planning department in the City of Cape Town has also expressed concern, pushing on various fronts for affordable rentals – both private and public – to be developed in well-located urban areas. A number of publicly owned and well-located sites have been identified for affordable rental housing by planners who are intent on reversing the processes of exclusion (the Two Rivers Urban Park, a project with no connection to Nairobi's development of a similar name, is one of the largest sites). While this process is supported by the housing departments in the City and Province, intent on reaching their rental housing targets, the custodians of this land, Provincial Transportation and Public Works, are less than excited about releasing assets at under market value, a drama marked by ongoing negotiation, unfulfilled memoranda of understanding (MOUs), court cases and interdepartmental hostility.

Gated suburban estates

The middle class and wealthy make up a small proportion of Cape Town's overall population; over 70 per cent of the population is unable to afford a mortgage on the basis of income (Cirolia, 2016). Despite this, gated suburban estates serve as an important force in the expansion of the city. Middle- to upper-income households, often first-time homebuyers looking for low-maintenance and safe options, hoping for the 'idyllic' suburban home and garden, yet unable to afford these types of property in high-end suburbs (such as the Central City, Constantia or the Atlantic Seaboard), opt for the new-build 'gated estates' (Morange *et al.*, 2012; Lemanski *et al.*, 2008). These estates are generally 'controlled access' developments, rather than completely privatized spaces; however, there are many variants (Lemanski *et al.*, 2008).

The rise in popularity of these developments in Cape Town is dated to the early 2000s (ibid.). These villages tend to be on the periphery of Cape Town, most notably in the northern suburbs (areas such as Durbanville and Tygerberg) – what Turok (2001) calls the 'northern drift' (Morange *et al.*, 2012). The Victoria and Albert Waterfront and the accompanying gated residential developments remains the only major projects of this kind in the Central City (Marks and Bezzoli, 2001).

Century Palace (among other gated estates in Century City) is a good example of this sort of development. Century City, a pseudo-Tuscan mega-project located on the N1 national highway, includes a large mall, smaller shopping areas, an office park (home to many call centres) and small amusement park (ibid.). The gated estates surrounding Century City, like Century Palace, are thus given value by the investments of Century City, an area which was previously grassy marshland. Many smaller projects draw on similar design and development principles – in short, building a whole suburban environment to create residential development.

As these developments become more common in and around Cape Town, the public planning sector, in particular, has taken on an anti-gating policy. The city planning department clearly recognizes the problems with gating space. However, the private developers increasingly find ways to circumvent it, arguing that their contribution to densification and mixed-use developments, in fact, overrides the negative social impacts of their developments (Didier *et al.*, 2012; Lemanski, 2006).

Cape Town reflections

Unlike Nairobi, active efforts on the part of the local government have had a huge impact on the spatial development of the city. Both RDP housing delivery and the redevelopment of the CBD have been led by the local government – albeit, with contestation between departments and spheres. In many ways, the regeneration process – coupled with perceived issues of security – have also had knock-on effects, the rise of peripheral gated and security estates being one of the implications of the lack of affordability for the middle class in well-located urban areas. However, the state is neither homogeneous nor has it had full control over development of the city. The rise of backyard shacks, the escalating unaffordability of well-located areas and the rise of gated estates have challenged the hegemony of state control and highlighted its fault lines.

Comparing Nairobi and Cape Town: contested and political nature of urban development?

The above patterns discussed in Nairobi and Cape Town do not seek to encapsulate all of the development in both cities. Instead, they seek to identify some of the key patterns emerging in both cities and the actors and pressures – regarding land, planning and infrastructure – which each involve. These narratives offer insights into accumulative geographical and political processes which result in substantial shifts, over time, in each city.

While there are many similarities and differences in the development patterns in Nairobi and Cape Town, both cases combat simplistic readings of African cities as *only* cases of slum urbanism. Both cities experience informality and elite enclaves; however, they do so in ways which are substantially more nuanced and variegated than often assumed. The city planning departments in both Nairobi and Cape Town are important actors in their respective cases, albeit in different ways. In both cases, city planning departments are in constant negotiation – struggling to claim their roles in the actual and grounded processes of city-making. This complexity reflects a 'negotiated city-making', rather than a simple narrative of planning's failure or success. This negotiated-ness sits in contrast to the assumed 'planning-leads-development' understanding of urban change.

In the case of Nairobi, this negotiation is most frequently between the planning department and small- and medium-scale developers. It reflects a more 'retroactive' attempt to shape or control development. In Nairobi, the local government is only just coming to grips with the implications of their devolved responsibility. The state has not played an aggressive role in shaping the urban fabric to date.

With rampant speculation and unchecked tenement, peri-urban and even gated development, the local government is moving to retroactively negotiate the urban fabric. In this context, the state is trying to make sense of the scaled need for very low-income and largely informal accommodation, on the one hand, and the making of a 'world-class African city' and the excited landing of global investment capital, on the other. Here, planning plays a largely reactive role, seeking to regain lost power in a patchwork and negotiated manner.

In the case of Cape Town, in contrast, the planning department negotiates more frequently with other state departments, reflecting conflicting operating logics and mandates among state departments and institutions. In Cape Town, efforts to plan a just and integrated city firmly contradict other departments and their mandates, be it to deliver housing, maintain their asset profiles, or respond to the demands of property rates payers. Simultaneously, planning efforts must grapple with the larger developers' interest in the supply of gated and anti-social housing estates and low-income RDP housing owners' desire to earn a livelihood out of their property. While the planning department has many planning instruments aimed at ensuring that a system of infrastructure provision and development regulation align, the routine demands of citizens (both rich and poor) and, even more importantly, other departments (such as human settlements or property disposal units), force the local government into a negotiation, with material impacts on the urban fabric of Cape Town.

These more complex narratives work to 'de-pathologize' the African city. The work of French planner and geographer Sylvy Jaglin is particularly useful in this regard. Jaglin (2014: 34) proposes:

> a radical change in perspective, taking as a starting point not the failure of urban services and the institutions responsible for their delivery, but the vitality and multiplicity of actual delivery systems which, despite policy announcements and reforms, and notwithstanding imported models, survive and contribute to the functioning of cities.

This change in orientation allows for an embrace of the variegated and negotiated nature of urban planning in Cape Town and Nairobi. It allows for the relaxation of the binary conventions placed on the boundaries of the 'the public' and 'the private' and a muddying of the normative assumptions about what the city-making roles of the municipality, the national government, the developer and even the household could be. It allows for the diversity, contestation and complexity which is fundamentally true to city-making to be exposed, understood and even valorized.

Notes

1 The Cape Town research has been conducted as part of the African Centre for Cities Sustainable Human Settlements Citylab, of which the author is the coordinator. The Nairobi research has been conducted as part of the African Centre for Cities

DFID-funded study of land value capture in three African cities. The author conducted and consolidated the Nairobi sections of this work.

2 See the websites and promotional videos for various projects, including Thika Greens (www.thikagreens.co.ke), Garden City (gardencity-nairobi.com) and Four Ways Junction (www.youtube.com/watch?v=2kOliuPGoIs).

References

Adams, D., and Tiesdell, S. (2010) Planners as market actors: rethinking state-market relations in land and property. *Planning Theory and Practice*, 11(2): 187–207.

Alonso, W. (1960) A theory of the urban land market. *Papers in Regional Science*, 6(1): 149–57.

Amin, A. (2014) Lively infrastructure. *Theory, Culture and Society*, 31(7–8): 137–61.

Amis, P. (1984) Squatters or tenants: the commercialization of unauthorized housing in Nairobi. *World Development*, 12(1): 87–96.

Avni, N., and Yiftachel, O. (2014) The new divided city? Planning and 'gray space' between global north-west and south-east. In S. Parnell and S. Oldfield (eds), *The Routledge Handbook on Cities of the Global South*. London: Routledge: 487–505.

Bekker, S., and Therboro, G. (2011) Introduction. In S. Bekker and G. Therboro (eds), *Capital Cities in Africa: Power and Powerlessness*. Dakar, Cape Town: CODESRIA, HSRC: 1–6.

Bremner, L. (2000) Reinventing the Johannesburg inner city. *Cities*, 17(3): 185–93.

Callies, D., and Grant, M. (1991) Paying for growth and planning gain: an Anglo-American comparison of development conditions, impact fees and development agreements. *The Urban Lawyer*, 23(2): 221–48.

Cirolia, L. R. (2016) Reframing the 'gap market': lessons and implications from Cape Town's gap market housing initiative. *Journal of Housing and the Built Environment*, 31(4): 621–34.

DFID (2015) Urban infrastructure in sub-Saharan Africa: harnessing land values, housing and transport. Report on Nairobi Case Study (Report No 1.8). Available from: www.africancentreforcities.net/wp-content/uploads/2015/09/DfID-Harnessing-Land-Values-Report-1.8-Nairobi-Case-Study-20150731.pdf. Accessed 5 September 2016.

Didier, S., Peyroux, E., and Morange, M. (2012) The spreading of the city improvement district model in Johannesburg and Cape Town: urban regeneration and the neoliberal agenda in South Africa. *International Journal of Urban and Regional Research*, 36(5): 915–35.

Ennis, F. (1997) Infrastructure provision, the negotiating process and the planner's role. *Urban Studies*, 34(12): 1935–54.

Estache, A. (2010) Infrastructure finance in developing countries: an overview. *EIB Papers*, 15(2): 60–88.

Franzsen, R. (2003) Property taxation within the Southern African Development Community (SADC): current status and future prospects of land value taxation, Botswana, Lesotho, Namibia, South Africa and Swaziland. Lincoln Institute of Land Policy, Working Paper (WP03RF1).

Gatabaki-Kamau, R., and Karirah-Gitau, S. (2004) Actors and interests: the development of informal an informal settlement in Nairobi, Kenya. In K. T. Hansen and M. Vaa (eds), *Reconsidering Informality: Perspectives from Urban Africa*. Upsala: Nordiska Afrikainstitutet: 158–75.

Gilbert, A. (2009) Extreme thinking about slums and slum dwellers: a critique. *SAIS Review*, 29(1): 35–48.

Graham, S. (2010) When infrastructures fail. In S. Graham (ed.), *Disrupted Cities*. New York: Routledge: 1–26.

Graham, S., and Marvin, S. (2001) *Splintering Urbanism: Networked Infrastructures, Technological Mobilities and the Urban Condition*. Abingdon: Psychology Press.

Grant, R. (2015) Sustainable African urban futures: stocktaking and critical reflection on proposed urban projects. *American Behavioral Scientist*, 59(3): 294–310.

Gulyani, S., and Talukdar, D. (2008) Slum real estate: the low-quality high-price puzzle in Nairobi's slum rental market and its implications for theory and practice. *World Development*, 36(10): 1916–37.

Guy, S., and Hanneberry, J. (2002) *Development and Developers: Perspectives on Property*. Oxford: Blackwell.

Harrison, P. (2006) Integrated development plans and third way politics. In U. Pillary, R. Tomlinson and J. du Toit (eds), *Democracy and Delivery: Urban Policy in South Africa*. Cape Town: HSRC: 186–207.

Harrison, P., Todes, A., and Watson, V. (2008) *Planning and Transformation: Lessons from the South African Experience*. London: Routledge.

Harvey, D. (2012) *Rebel Cities: From the Right to the City to the Urban Revolution*. London: Verso.

Hass Property Index (2009) *Quarter Four Report 2009*. Available from: https://www.hassconsult.co.ke/images/Q4.9%20Report.pdf. Accessed 24 March 2017.

Healey, P., and Barrett, S. (1990) Structure and agency in land and property development processes: some ideas for research. *Urban Studies*, 27(1): 89–103.

Huchzermeyer, M. (2008) Slum upgrading in Nairobi within the housing and basic services market: a housing rights concern. *Journal of Asian and African Studies*, 43(1):19–39.

Huchzermeyer, M. (2011) *Tenement Cities: From 19th Century Berlin to 21st Century Nairobi*. Trenton, NJ: Africa World Press.

Ingram, G., Lui, Z., and Brandt, K. (2013) Metropolitan infrastructure and capital finance. In R. W. Bahl, J. F. Linn, and D. L. Wetzel (eds), *Financing Metropolitan Governments in Developing Countries*. Cambridge, MA: Lincoln Institute of Land Policy: 339–66.

Jaglin, S. (2014) Rethinking urban heterogeneity. In S. Parnell and S. Oldfield (eds), *The Routledge Handbook on Cities of the Global South*. London: Routledge: 434–46.

Kihato, M. (2012) Infrastructure and housing finance: exploring the issues in Africa. Available from: www.housingfinanceafrica.org/document/infrastructure-and-housing-finance-exploring-the-issues-in-africa/. Accessed 5 September 2016.

Kihoro, M. W. (2015) Factors affecting performance of projects in the construction industry in Kenya: a survey of gated communities in Nairobi county. *Strategic Journal of Business and Change Management*, 2(2): 37–66.

Lemanski, C. (2006) Desegregation and integration as linked or distinct? Evidence from a previously 'white' suburb in post-apartheid Cape Town. *International Journal of Urban and Regional Research*, 30(3): 564–86.

Lemanski, C., and Oldfield, S. (2009) The parallel claims of gated communities and land invasions in a southern city: polarised state responses. *Environment and Planning A*, 41(3): 634–48.

Lemanski, C., Landman, K., and Durington, M. (2008) Divergent and similar experiences of 'gating' in South Africa: Johannesburg, Durban and Cape Town. *Urban Forum*, 19(2): 133–58.

Mabin, A., and Smit, D. (1997) Reconstructing South Africa's cities? The making of urban planning 1900–2000. *Planning Perspectives*, 12(2): 193–223.

Marks, R., and Bezzoli, M. (2001) Palaces of desire: Century City, Cape Town and the ambiguities of development. *Urban Forum*, 12(1): 27–48.

Martin, L. H. (2006) Safari in the age of Kenyatta. *The Hemingway Review*, 25(2): 101–6.

McFarlane, C., and Robinson, J. (2012) Introduction: experiments in comparative urbanism. *Urban Geography*, 33(6): 765–73.

McKinsey and Company (2010) Lions on the move: the progress and potential of African economies. Available from: www.mckinsey.com/global-themes/middle-east-and-africa/lions-on-the-move. Accessed 5 September 2016.

McKinsey and Company (2012) The rise of the African consumer. Available from: www.mckinsey.com/global-locations/africa/south-africa/en/rise-of-the-african-consumer. Accessed 5 September 2016.

Medard, C. (2010) City planning in Nairobi. In H. Charlton-Bigol and D. Rodriguez Torres (eds), *Nairobi Today: The Paradox of a Fragmented City*. Tanzania: Kathala: 25–60.

Miraftab, F. (2005) Insurgency and spaces of active citizenship: the story of Western Cape anti-eviction campaign in South Africa. *Journal of Planning Education and Research*, 25(2): 200–17.

Morange, M., Folio, F., Peyroux, E., and Viviet, J. (2012) The spread of a transnational model: 'gated communities' in three southern African cities (Cape Town, Maputo and Windhoek). *International Journal of Urban and Regional Research*, 36(5): 890–914.

Murray, M. (2008) *Taming the Disorderly City: The Spatial Landscape of Johannesburg after Apartheid*. New York: Cornell University Press.

Mwau, C.B. (2013) *The Gradual Decline of the 'Zinc Age': Tenements For Nairobi's Low Income Population*. Cape Town: University of Cape Town.

Myers, G. (2014) From expected to unexpected comparisons: changing the flows of ideas about cities in a postcolonial urban world. *Singapore Journal of Tropical Geography*, 35(1): 104–18.

Myers, G. (2015) A world-class city-region? Envisioning the Nairobi of 2030. *American Behavioral Scientist*, 59(3): 328–46.

Nairobi City County (2014) Integrated urban development masterplan for the city of Nairobi. Available from: http://citymasterplan.nairobi.go.ke/index.php/latest-news/79-niuplan. Accessed 5 September 2015.

Napier, M., Berrisford, S., Kihato, C. W., McGaffin, R., and Royston, L. (2013) *Trading Places: Accessing Land in African Cities*. Cape Town: African Minds.

Oldfield, S. (2014) Critical urbanims. In S. Parnell and S. Oldfield (eds), *The Routledge Handbook on Cities of the Global South*. London: Routledge: 7–8.

Owuor, S., and Mbatia, T. (2011) Nairobi. In S. Bekker and G. Therboro (eds), *Capital Cities in Africa: Power and Powerlessness*. Dakar, Cape Town: CODESRIA, HSRC: 120–40.

Parnell, S., and Pieterse, E. (2014) Africa's urban revolution in context. In S. Parnell and E. Pieterse (eds), *Africa's Urban Revolution*: 1–17.

Parsons, T. (2013) 'Kibra is our blood': the Sudanese military legacy in Nairobi's Kibera location, 1902–1968. *The International Journal of African Historical Studies*, 30(1): 87–122.

Pieterse, E. (2008) *City Futures: Confronting the Crisis of Urban Development*. London: Zed.

Pieterse, E., and Cirolia, L. R. (2016) South Africa's emerging national urban policy and upgrading agenda. In L. R. Cirolia, T. Görgens, and M. van Donk (eds), *Pursuing a*

Partnership Based Approach to Incremental Informal Settlement Upgrading in South Africa. Cape Town: University of Cape Town Press: 553–65.

Pirie, G. (2007) 'Reanimating a comatose goddess': reconfiguring Central Cape Town. *Urban Forum*, 18(3): 125–51.

Rakodi, C. (2007) Land for housing in African cities: are informal delivery systemes instituionally robust and pro-poor? *Global Urban Development*, 3(1): 1–11.

Robinson, J. (2006) *Ordinary Cities: Between Modernity and Development*. London: Routledge.

Robinson, J. (2016) Thinking cities through elsewhere: comparative tactics for a more global urban studies. *Progress in Human Geography*, 40(1): 3–29.

Roy, A. (2005) Urban informality: towards an epistemology of planning. *Journal of the American Planning Association*, 71(2): 147–58.

Simone, A. (2010) *City Life from Jakarta to Dakar: Movements at the Crossroads*. London: Routledge.

Smit, W. (2015) *Transforming Cities: Analysing the Recontextualisation of Discourses of the Urban in Post-Apartheid Cape Town*. Cape Town: University of Cape Town.

Taylor, W. E. (2004) Property rights – and responsibilities? The case of Kenya. *Habitat International*, 28(2): 275–87.

Thorn, J., and Oldfield, S. (2011) A politics of land occupation: state practice and everyday mobilization in Zille Raine Heights, Cape Town. *Journal of Asian and African Studies*, 46(5): 518–30.

Turok, I. (2001) Persistent polarisation post-apartheid? Progress towards urban integration in Cape Town. *Urban Studies*, 38(13): 2349–77.

Visser, G., and Kotze, N. (2008) The state and new-build gentrification in Central Cape Town, South Africa. *Urban Studies*, 45(12): 2565–93.

Watson, V. (2003) *Change and Continuity in Planning: Metropolitan Planning in Cape Town Under Political Transition*. London: Routledge.

Watson, V. (2014) African urban fantasies: dreams or nightmares? *Environment and Urbanization*, 26(1): 215–31.

Wilson, S. (2011) Planning for inclusion in South Africa: the state's duty to prevent homelessness and the potential of 'meaningful engagement'. *Urban Forum*, 22(3): 265–82.

Yahya, S. (2002) Community land trusts and other tenure innovations in Kenya. In G. Payne (ed.), *Land, Rights and Innovation, Improving Tenure Security for the Urban Poor*. London: ITDG: 233–63.

3 Ordinary urban geopolitics
Contrasting Jerusalem and Stockholm

Jonathan Rokem

Introduction

This chapter joins the recent topical debate within urban geography and planning studies concerning the Global North's declining dominance in the production of urban theory and the need to move beyond methodological regionalism and incommensurability in urban studies research (Parnell and Robinson, 2012; Sheppard *et al.*, 2013; Watson, 2013, 2014; Peck, 2015). Within this discussion, a related process is the tightening gap between neoliberal infused socio-economic segregation and more extreme ethnic division in a growing number of cities worldwide. The proposition put forward is that through analysing urban geopolitics and planning within two defined case studies we can establish a contrasting comparative investigation of urban difference (McFarlane and Robinson, 2012).

In this respect, the two cases were chosen since they hold contrasting historical and geopolitical welfare settings with the urban scale, holding significant lessons. In Stockholm, government failure since the 1990s to tackle urban segregation has resulted in a growing spatial and social disconnection of immigrants placed in outer suburbs, initiating a rising separation of ethnic minority populations (Anderson, 2010). In Jerusalem, an active ethno-national conflict and the ongoing dilapidation and lack of development in Palestinian Arab areas have caused extreme spatial and social segregation. The findings suggest national policies have enduring consequences on the exclusion of minority groups, as the local case of *Al-Isawiyyah* will demonstrate. This is further illustrated in the case of *Fittja*, an outer suburb in the southern fringes of Stockholm, where the well-intended anti-segregation policies cause growing ethnic divisions, pointing us towards the need to rethink the nature and scope of urban geopolitics in different contested cities.

Jerusalem and Stockholm is an unusual pair to compare, demanding a short explanation, because they do not hold the same history or social and spatial politics. As I show in this chapter, there is a larger spectrum of state-led planning apparatuses producing urban segregation involved in both, and it is significant for critical urban theory to consider this more generally by looking at the specific national and local manifestations of this worldwide process (Porter, 2014: 388–9).

While in no way denying or condoning the role of the nation state in imposing the partisan segregation and inequality characterizing Jerusalem or Stockholm, it can be instructive to place it in a comparative context beyond the processes of ethno-national-led segregation, allowing us to see that there are also other forces in action, such as socio-economic states (Rokem, 2016).

Methodologically, the chapter grounds the analysis within three themes, examining: (1) housing and development, (2) mobility and transport and (3) local government and civil society. The empirical materials for this research were gathered through a synthesis of the author's fieldwork conducting 60 interviews with municipal planners and community activists in Jerusalem and Stockholm (2011–13) combined with a range of secondary material including professional reports, planning documents and newspaper articles. The chapter opens with a brief critical overview of the scope and limits of comparative urban theory and urban geopolitical research. Next, a brief description of the two cities is given, followed by an examination of three themes constructing a comparative conversation across and within Jerusalem and Stockholm. The conclusion ties all of these together, suggesting that it is timely to start learning from, and compare across, radically different cities.

Why compare urban geopolitical difference?

This chapter joins the recent call within critical urban studies and comparative urbanism to "move beyond many of the ethnocentric assumptions currently embedded in urban theory" (McFarlane and Robinson, 2012: 767; Robinson, 2002; Roy, 2009, 2014). It is critical to grasp the blurring of pre-proposed urban labels and concepts within the recent wider debate about the repositioning of urban theory, from a few dominant cases in the Global North (Parnell and Robinson, 2012; Watson, 2013, 2014; Sheppard *et al.*, 2013; Peck, 2015) to a wider assessment of the 'world of cities' (Robinson, 2011, 2014) in an ever more fractured urban geopolitical reality.

In recent adaptations to classic geopolitics (Dalby, 1990; Agnew, 2003) there is a growing interest in the urban scale shifting from the classic geopolitical focuses on national actors and governmental policies to different local contested urban sites (Graham, 2004, 2010; Sidaway, 2009; Fregonese, 2009, 2012). Urban geopolitics has focused on the contested nature of ethno-nationalism at the urban scale. This has been well documented over the past few decades. To mention but a few cases, Belfast, Jerusalem, Johannesburg, Sarajevo, Baghdad, Beirut, Kirkuk and Mostar are repeatedly cited as purportedly manifesting extreme ethno-national divisions emanating from the 'contestation of the nation state' (Anderson, 2010; Gaffikin and Morrissey, 2011). Within this literature, urban transformations have been frequently analysed through urban planning and ethno-national politics (see Hepburn, 2004; Calame and Charlesworth, 2009; Bollens, 1998, 2007, 2012; Pullan and Baillie, 2013).

One such example, on the one hand, is the case of East Jerusalem, where the state is actively restricting Palestinian Arabs from building new houses in their neighbourhoods. In Sweden, on the other hand, one of the main issues is that the Swedish welfare state was conceived in the first half of the twentieth century, at a time when there was a relatively hegemonic population. It is currently struggling to cope with its growing ethnic national diversity. Subsequently, the well-intended immigration and social housing policies placing immigrants and asylum seekers in remote outer suburbs is leading to divisions between the latter and the native Swedish population – contrasting with East Jerusalem, where the native Palestinians are the minority population.

One of the explanations for the lack of incorporation of minority groups in both cases is an international geopolitical system that holds an ideal of hegemonic nation states. In reality we know that there are few cases where *nation* and *state* are congruent, a situation which might suggest that cities not currently affected by overt ethno-national conflict, but rather extreme ethnic segregation from diverse ethnic and national origins (Stockholm in the current study), may, sometime in the future, turn to a more radical form of ethno-national conflict (such as in Jerusalem). With this in mind, I propose socio-economic state and its relation to ethnicity as the two main defining sequential segregation causes. In the next section, Jerusalem's urban scale is reviewed followed by that of Stockholm and the local cases of Fittja and Al-Isawiyyah are brought into the comparative discussion.

To start establishing a comparative conversation of what we can learn from different urban contexts (McFarlane and Robinson, 2012: 766), I suggest that there is a need to rethink our current theoretical categories and labels, based on empirical research in two specific cities representing radically different visions and division patterns. Such a step could contribute to one of the long-standing questions at the core of urban theoretical enquiry concerning the validity of singular cases (cities) in the creation of a general urban theory? "[C]an [we] group all cities together as a common class of phenomena? Or must we divide them into several different and incommensurable classes, and, in the extreme case, into as many classes as there are individual cities?" (Scott and Stroper, 2014: 8).

This highlights a theoretical and methodological risk: first, of collapsing into a deterministic proposition trap that all or most of today's cities are undergoing similar changes; or, second, that they are all unique and incommensurable. It is important to clarify that neither extremes are the aim of the current chapter. Rather, the aim follows Peck's (2015) observation: "The ongoing work of remaking of urban theory must occur across cases, which means confronting and problematizing substantive connectivity, recurrent processes and relational power relations, in addition to documenting difference, in a 'contrastive' manner, between cities" (Peck, 2015: 162–3).

In other words, this chapter suggests we need to find a middle ground by carefully mapping and contrasting the proposed three themes defining urban geopolitics in two different cities to reveal some preliminary observations of what we can learn from cities previously deemed to be fundamentally incommensurable in urban comparative research (Robinson, 2011: 5).

Jerusalem: ethno-national and socio-economic segregation

The contested nature of planning in Jerusalem is widely recognized – and, indeed, the city is firmly placed in the literature on 'divided cities' (Bollens, 2012; Dumper, 2014; Shlay and Rosen, 2015) – see Rokem and Allegra (2016) for a review. Jerusalem is the largest and poorest city in Israel today. According to the Israel Central Bureau of Statistics (ICBS, 2014), at the end of 2013 the population of Jerusalem numbered 816,000. The Jewish population totalled 515,000 (63 per cent), the Arab (Muslim and Christian) and other (non-Jewish) population totalled 301,000 (37 per cent). Several factors distinguish Jerusalem from other cities. First, it is an important religious centre for three of the world's monotheistic religions. Second, it is claimed as the national capital by two nations, placing it in the vortex of the Israeli–Palestinian conflict, and, third, it is not acknowledged as the official capital of Israel by the UN and most of the world's nation states.

One of the main reasons for the tight control of planning in Israel, especially in Jerusalem, is the ongoing Palestinian–Israeli conflict. The 1948 war[1] ended with the city physically divided between two states, Jordan in the east and Israel to the west. The 1967 war[2] between Israel and its Arab neighbours was a significant spatial turning point in Israel's geopolitical condition – with the annexation of the Golan Heights, the Gaza Strip and the West Bank, including East Jerusalem – with the Israeli state declaring the city as its united capital (Bollens, 2000; Hasson, 2007; Dumper, 2014).

Since 1967, the Palestinian inhabitants of East Jerusalem have not been recognized as Israeli citizens and receive limited residency rights, which have been constantly eroded over time (UNCTAD, 2013). Israel's planning policies have been consistently aimed at strengthening Israel's control of East Jerusalem while weakening the rights of Palestinians (Khamaisi, 2002, 2010). Furthermore, the Israeli Ministry of Interior and the Jerusalem Municipality placed a strict development ban, forbidding almost any new construction in Palestinian neighbourhoods (Rokem, 2013). Moreover, plans for developments over a certain size have to receive the Ministry of Interior-controlled regional planning committee's approval (Planning Administration, 2013: 122), thus placing the political power to approve large-scale plans in state hands.

The underlying principle of Israeli planning policy in Jerusalem is to establish a large, unified city with a dominant Jewish majority. The *Jerusalem Master Plan 2000* is the newest plan for the entire city and was adopted in 2007 by the municipality. Although the Master Plan remains the main planning guidance policy document, it has not received statutory approval by the Israeli government – enabling its selective use by the authorities. The plan proposes a population ratio of 60 per cent Jewish to 40 per cent Palestinian, maintaining this demographic balance in the future. This is an alteration of the former 70 per cent to 30 per cent ratio and reveals the failure of attempts to control the demographic balance through planning. For the first time since 1967, the current *Jerusalem Master Plan 2000* relates to the development and

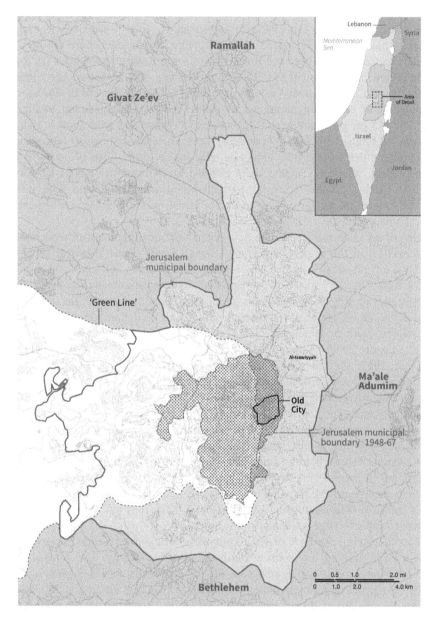

Figure 3.1 Jerusalem's borders and boundaries
Source: Author 2016

planning needs of the entire city's population. The Director of the Jerusalem
Municipality Planning Department expressed in an interview his satisfaction
with the Master Plan's improving of the transparency of planning policy in
Jerusalem.

Since the Master-plan 2000 was approved for authorization by the local planning committee… planning policy has been much clearer… even though it has not received statuary approval from the district committee it makes it much easier to decide where and how to develop.

Interview, Director Jerusalem Municipality Planning
Department, May 2013

Notwithstanding the planning transparency affirmed in the statement above, the *Jerusalem Master Plan 2000*'s lack of either statuary approval or determination of detailed land use means that, in reality, it is selectively implemented, mainly targeting Jewish areas. In order to develop new areas contained in the plan there is a need for a detailed local plan, which regulates the type of land usage to receive planning permission for a building permit. But, for the majority of the Palestinian areas in East Jerusalem, there are no legally approved outline plans and, consequently, these areas remain neglected. This fact is rarely mentioned in official municipal planning policy, but was confirmed in an interview with a senior Jerusalem Municipal Planner: "No larger-scale plans for Palestinian neighbourhoods have been approved by the municipality; they are all frozen at the moment… the reason is political and has nothing to do with planning" (Interview, Senior Planner Jerusalem Municipality, March 2013).

The unequal funding of urban planning and construction projects between the Palestinian and Jewish parts of Jerusalem has resulted in a segregated city split into two distinct growth poles. Contrasting political contexts are observed next in Stockholm's urban reality.

Stockholm: ethnic segregation and the declining welfare state

Stockholm, founded in 1252, is the capital of Sweden. It is a rapidly expanding urban area located around Lake Mälaren, with 897,700 inhabitants at the end of 2013 and a total foreign-born population of 30.7 per cent (Stockholm Office of Research and Statistics, 2015). Since the 1980s Stockholm has experienced growing segregation and division emerging from the large number of migrant labourers and asylum seekers, coming mainly from Eastern Europe, Africa and the Middle East (Musterd, 2005: 333). Spatial segregation in Stockholm has a clear geographical pattern (Andersson, 2007: 66). The central areas and inner city suburbs have the least poverty and fewest immigrants. The concentration of immigrants in small areas has created neighbourhoods where non-immigrant Swedish citizens choose not to live or visit.

In the Swedish government's *Divided Cities Report* (Storstadskommittén, 1997), five areas from the larger Stockholm region were identified as suffering from extremely low income: Fittja, Vårby, Rinkeby, Tensta and Hovsjö. All five areas populated mainly by immigrants and located in Million Homes Programme areas are part of a grand project instated by the post-Second World War Swedish government to build a million new homes within a decade. They were constructed between 1965 and 1974, mainly in new self-contained modernist urban

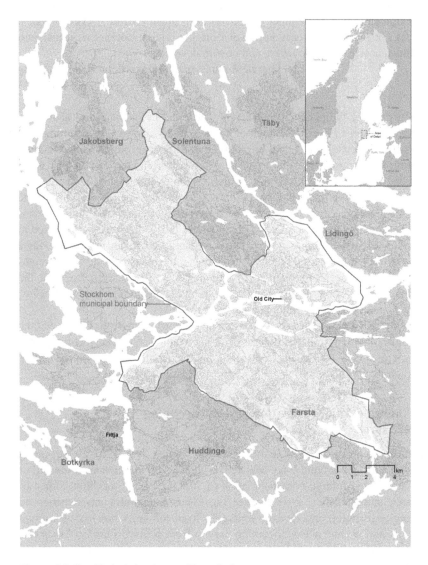

Figure 3.2 Stockholm's borders and boundaries
Source: Author 2016

developments in the outer periphery of Sweden's larger cities (Hall and Vidén, 2005: 301). In the Stockholm area new neighbourhoods were built in the northern parts around Järvafället, and on the southern fringes of the city in Skärholmen and Botkyrka Municipality. The growing segregation of Stockholm's Million Homes Programme outer suburbs was reported during an interview with an urban planner working in Botkyrka: "Immigrant neighbourhoods in Botkyrka and elsewhere in

Greater Stockholm have a negative connotation for the Swedish public and media.… It is a wider issue and needs to be changed by strengthening the local identity and connection to Swedish society" (Interview, Botkyrka Municipality Urban Planner, August 2012).

The Swedish government's immigration policies classify all newcomers to Sweden as immigrants (refugees and asylum seekers from poorer countries). This unified classification fails to recognize the different cultural and ethnic background and does not grant individual representation or voice. Placing them in isolated outer suburbs creates an even stronger feeling of estrangement from the Swedish majority population.

Botkyrka was chosen as the focus for the Swedish case study since it is an extreme illustration of ethnic segregation, having the second highest concentration of immigrants in Sweden (Petersson, 2011: 19). In my fieldwork, the main focus was the denser, more urban area of 'North Botkyrka', consisting of three Million Homes Programme areas (Fittja, Alby and Hallunda-Norsborg). At the end of 2013, Botkyrka's total population was 87,600, with 41,000 in the much smaller urban northern area and 46,600 living the more rural south. North Botkyrka's immigrant population was 91.6 per cent in 2013 (Botkyrka Statistics, 2014) and the Swedish foreign-born national average was 7.2 per cent (SCBS, 2014),[3] illuminating the urban geopolitical dimension of the problem.

Thematic anchor points of urban difference

This part of the chapter is going to reflect comparatively across the thematic axis focusing on dwelling, mobility and complex governance structures, jointly composing a set of interrelated and overlapping patterns through which to uncover local conditions and voices across cases. I expose how *plural causal factors* at the same approximate point in time produce comparable outcomes (Pickvance, 1986, 2001). In other words, I suggest questioning the long-standing methodological regionalism and incommensurability moving beyond the incomparability of cities from diverse geopolitical settings.

Housing and development in ethnic minority areas

Uncontrolled sprawl and (lack of) *housing* and development in East Jerusalem Palestinian areas and continued absence of investment and deteriorating housing conditions in Stockholm's outer suburbs can be determined as a main *housing and development anchor point*, reinforcing ethnic segregation and deprivation. In both instances this process is led by the state and further reinforced by local municipal government. On the one hand, in most of East Jerusalem's Palestinian neighbourhoods there has been no official outline plan developed. The result is construction of *illegal* housing by local residents under constant threat of demolition by Israeli authorities (Braverman, 2006).

In Sweden, on the other hand, when new immigrants arrive, whether voluntarily or to escape harsh domestic conditions (as in the case of refugees and asylum seekers), the Swedish government places them where there is vacant public housing, mostly in suburban Million Homes Programme neighbourhoods on the edges of the large cities. However, the success of government policies in changing the local area was further questioned in the interview with Fittja's Community Development Officer, who criticized their long-term impact: "[The Swedish government's initiatives] are like taking pain killers, but not solving the real underlying illness" (Interview, Community Development Officer, Fittja, August 2013).

In relation to housing, both cases have tight control over planning and development. However, there are stark differences in the causes of the regulatory constriction. While in Stockholm most of the immigrants live in publicly owned housing managed by the Botkyrka Housing Association, in Jerusalem restrictions on housing and development is officially used by the municipality to label any new construction 'illegal' as it lacks planning permission. The other side of the story is an overall policy to keep the demographic balance by obstructing mechanism rather than a positive tool intended to benefit the local Palestinian Arab population. This is reflected in the Jerusalem Municipality's politically motivated agenda to hinder development of a local Al-Isawiyyah outline plan, further voiced in the interview with a Palestinian community activist:

> How can we expect the municipality to promote a plan when the basic needs of the residents are not taken care of… We have no sidewalks for the children to walk home safely, there are no playgrounds… garbage is only removed sporadically, creating a health hazard.
>
> Interview, Al-Isawiyyah community activist, May 2013

In East Jerusalem there is an *off-the-record* political agenda to hamper development for Arab Palestinians (Rokem, 2016). As previously mentioned, the lack of available housing in Palestinian areas is a growing concern and cause for local resistance. In Sweden, in contrast, public housing deterioration is part of the declining welfare state and lack of finance for urban renewal. The lack of public funding to regenerate depleted areas such as Fittja result in local opposition by immigrant-led social movements against the selling of public housing to private developers to fund urban renewal.

Both local neighbourhoods contain marginalized groups, and are run down and neglected as a result of state actions stemming from different structural causes. Housing segregation alone is not necessarily a problem, but overcoming a combination of segregation and immobility becomes a challenge as shown in the next anchor point.

Mobility and transport

Public transport and personal mobility have critical implications for access to employment and other basic needs, especially in the case of marginalized minority communities. In Jerusalem, two separate public transport networks are used by Palestinians and Israelis, each having its own separate central bus station. The relatively new Light Railway Transit (LRT) modifies this to some extent by linking the East and West Central Bus Stations and serving both Palestinian and Israeli areas in the northern parts of the city (see Rokem and Vaughan, 2017). In Stockholm, this spatial and social segregation of immigrants in a limited number of outer suburbs reinforces the differentiated public transport system, especially subway lines ending in Million Homes Programme immigrant suburbs.

On the local scale there are numerous differences. Al-Isawiyyah is not connected to the new LRT network and has one local bus operated by the Palestinian bus network (Line No. 1) connecting it to East Jerusalem's central bus station by the Old City. Pedestrian movement to neighbouring areas is geographically proximate; there are indications that such access is becoming restricted. A community activist from the French Hill stated: "In general the area [Al-Isawiyyah] is regarded as a *no-go zone* by the Jewish population and has been perceived as a threat by nearby Jewish neighborhoods." In response to this perceived threat, a 'separation ditch' was built by the Jerusalem Municipality, "creating a new barrier to movement between Al-Isawiyyah and its nearest Jewish neighborhood, French Hill" (Haaretz, 2013).

In Stockholm, no physical barriers have been erected to date. However, there are several indicators for spatial and social segregation in the job market and public space (see Legeby, 2013). The most accessible public transport link from central Stockholm to Fittja is the Red Line *tunnelbana* (subway), built in the 1970 as part of the Million Homes Programme. In an interview with a researcher based in Fittja's Multicultural Centre (MKC), the difference in use of public transport between the residents of Fittja and South Botkyrka was described: "the slower *tunnelbana* serves the Fittja residents and the faster suburban train serves South Botkyrka's more rural *ethnic Swedish* population", resulting in relative absence of co-presence by immigrants and ethnic Swedes on the same public transport link. Pedestrian movement in Fittja was more prominent in the centre, where people walk to the *tunnelbana* stop and local shopping area. In an interview with the Botkyrka Municipal Strategic Planner, her own observation was that, "[t]he further away you get from the centre [of Fittja] relatively large empty spaces mean that the thought of walking there after dark is less appealing" (Interview, Botkyrka Municipality Urban Planner, August 2012).

Fittja is physically separated from the closest neighbouring communities by road and water barriers, making walking to the nearest quarter less probable. While Al-Isawiyyah is spatially more integrated, it is being separated from the surrounding urban fabric. Major changes have been a recently dug 'separation/ security ditch' and erection of security and surveillance infrastructure by the nearby Hadassah Hospital and the Hebrew University campus. The separation and limitations on mobility, in both cases, indicates diverging degrees of ethnic separation from the dominant majority culture. In the case of Jerusalem there are two separate public transport systems with a relatively new shared co-presence on the new LRT link; however, even in Sweden, there is a growing separation of movement between Botkyrka's northern and southern parts and lack of co-presence on the public transport link to central Stockholm. The third anchor point; *local governance and civil society* relates to the former two points: *housing and development* and *public transport and mobility*. Both stem from government priorities and lack of funding, which, in Jeruaslem and Stockholm, are determined at the national level, filtered down to the urban and neighbourhood scales and opposed by civil society.

Local governance and civil society

In both cases a dominant governance anchor point is the existence of top-down planning control and restrictions alongside a lack of connection to local communities' tangible needs and a different yet significant presence of civil society. In Stockholm this is reflected in the policy of labelling all newcomers to Sweden as *ethnic others* or immigrants, allocating them public housing in areas where they become spatially and socially isolated. In Jerusalem the lack of positive municipality involvement and focus on restricting development and housing demolitions has given room for civil society- and community-led activity, struggling to change the current reality. In contrast, Sweden in general has had a lack of civil society activity up until recently, with a strong presence of municipality officers. In the last decade this has shifted and new grassroots immigrant-based movements have mobilized as a result of a weakening welfare state. One such recent movement in Stockholm is *Megafonen*,[4] a grassroots civil society group known for actively supporting a mobilization against the Swedish authorities opposing the sale of rental flats to private developers to fund urban regeneration in Million Homes areas.

In Israel local municipalities have limited planning and development powers. They are responsible for approving detailed construction plans and granting building permits. In the political reality of East Jerusalem, where there is an agenda to hamper development, the local communities' mobilization on a site-based scale is the only hope for change. In Sweden, however, each municipality has its own master plan. The municipal authorities at the local scale are also responsible for planning approval and construction of the detailed plans (Lundström *et al.*, 2013). But the local picture is more complex than presented in the narrative of official policy reports. This is reflected by a Botkyrka municipal planner involved in the drafting of the Fittja Master Plan disclosing the lack of *hands-on* experience with the local population and their detachment from the daily conditions on the ground.

I have undertaken follow-up consultations and amendments with the municipality project group about Fittja's spatial Master Plan, 'Future Fittja': "I visited Fittja a few times and took part in one of the public consultations... I can say I know the place from the planning process but not so much the residents" (Interview, Botkyrka Municipality Urban Planner, August 2012).

The Swedish welfare state has traditionally seen its role as being to supply the entire population's socio-economic needs. There is a growing void between the well-intended municipal agenda and the perception by the local diverse ethnic minority immigrant population. One such example is the local residents' disbelief in the Fittja Master Plan participation process. This point was brought up in an interview with the Botkyrka Municipal Urban Planner leading the process, revealing the lack of trust in the authorities: "We don't want to say what we want again, we want to see concrete results" (Interview, Botkyrka Municipality Urban Planner, August 2012). This gives a glimpse of the stark contrast between the strong positive interest of Botkyrka Municipality in regenerating Fittja, and the local community's lack of interest and disbelief in the process. In Jerusalem, in contrast, there is a strong presence of civil society acting as mediator and representing minority populations in receiving planning permission from the authorities. In the Al-Isawiyyah case, this void was filled by the activity of the NGO Bimkom (Planners for Planning Rights). The Bimkom Project Manager leading the planning process stated in her interview:

> Bimkom decided to shift from their traditional role of critiquing planning injustice and violations of human rights of minorities in Israel, to actively work with the local Al-Isawiyyah community to develop a Master-Plan... the main aim was to cater for the future local population needs, especially the urgent lack of adequate planning conditions and threat of widespread housing demolitions.
>
> Interview, Bimkom Project Manager, July 2012

The Al-Isawiyyah local community believed in actively promoting better conditions as the only hope for change, but the struggle against the Jerusalem Municipality's politically motivated bureaucracy doomed it to failure. The *three anchor points* have shown extreme ethnic segregation in cities is embedded in complex political structures and socio-spatial patterns intertwined in different housing, mobility and local governance mechanisms, contradicting the official political agenda and received wisdom in each city. The growing division and mixing of ethnic minority populations challenges, on the one hand, the fragmentary segregation of communities in ethno-national contested cities, while, on the other hand, suggesting cities without overt ethno-national conflict are experiencing growing contestation and division stemming from different structural causes, yet holding potential promising comparative value.

Conclusion

The research suggests we need to open up our comparative imagination and develop a more relational *ordinary cities* framework (Robinson, 2006). In both Jerusalem and Stockholm, ethnic minority populations are informal in the sense that they are outside the dominant culture (Gaffikin and Perry, 2012: 712). The site visits and interviews with community activists and urban planners reveal political motives at different institutional scales that dominate planning decisions and their implementation by municipality planners, affecting local communities' overall individual and collective prospects. This was further reflected in the three *themes* demonstrating how housing conditions, daily movement, local governance and civil society actions produce different patterns in both cases, containing ethnic segregation stemming from diverse causes (Pickvance, 1986, 2001) with growing dissent and resistance towards the majority culture.

This chapter has endeavored to expose how ethno-national conflict in Jerusalem, can be perceived to have an enduring, totalizing nature that distinguishes it from other forms of conflict in an international system where the ideal of the nation state is hegemonic. Thus, the suggestion that Stockholm is not currently affected by overt ethno-national conflict but by ethnic segregation from diverse national origins does not hold up against the growing division and lack of integration of minority groups with the majority Swedish society.

As outlined throughout the text, extreme ethnic segregation in cities is a complex reality embedded in political structures – producing, in some cases, the opposite result to the original political agenda. On the one hand, this suggests that what are commonly labelled as ethno-national divided cities have a tendency for populations to mix, as the case of Jerusalem revealed. On the other hand, what has been perceived as well-balanced and sustainable urban development results in growing contestation and division of ethnic minority populations. As I have illustrated in the Stockholm case, the nation state's good intentions for immigrants, with housing and welfare assistance, has resulted in a growing division and lack of integration with the majority Swedish society. This brings us to questions about the need for a more nuanced understanding of the value of comparing urban difference rather than a continued focus on most similar cases in comparative urban studies, suggesting there is a growing need to rethink *labels* and *concepts* attributed to cities and neighbourhoods to better conceptualize and adapt policy and practice towards ethnic minorities and migrants in an ever more fractured urban geopolitical reality.

Acknowledgments and funding

Special thanks to Laura Vaughan and Margo Huxley for their insightful remarks on earlier versions of this text. This field research was supported by a fellowship (2012–13) from the French Research Institute, Jerusalem (CRFJ-CNRS).

Notes

1 'War of Independence' (Israeli name) or 'al-Nakbah' – the Disaster (Palestinian name); to simplify, the common term '1948 war' is used.
2 The 1967 Six Day War between Israel and its Arab neighbours ended in the occupation by Israel of the Sinai Peninsula, West Bank, Gaza Strip and Golan Heights.
3 Swedish Central Bureau of Statistics.
4 See www.megafonen.se.

References

Abu Lughod, J. (2007) The challenge of comparative case studies. *City*, 11(3): 399–404.

Agnew, J. (2003) *Geopolitics: Re-visioning World Politics*, 2nd edn. London: Routledge.

Anderson, J. (2010) *Democracy, Territoriality and Ethno-National Conflict: A Framework for Studying Ethno-Nationally Divided Cities*. Paper no.18. www.conflictincities.org. Accessed May 2015.

Andersson, R. (2002) Mobility in Fittja, an analysis of migration, segregation and big city politics. In I. Ramberg and O. Pripp (eds), *Fittja: The World and the Everyday*. Botkyrka: Botkyrka Multicultural Centre (Swedish).

Andersson, R. (2007) Ethnic Residential Segregation and Integration Processes in Sweden. In K. Schonwolder (ed.), *Residential Segregation and the Integration of Immigrants: Britain, the Netherlands and Sweden* [PDF]. Berlin: Social Science Research Center Berlin. Available from: http://uu.divaportal.org/smash/get/diva2:39365/FULLTEXT01: 61–90. Accessed May 2015.

Bimkom (2013) *Survey of the Palestinian Neighbourhoods in East Jerusalem* [PDF]. Jerusalem: Bimkom. Available from: www.bimkom.org/eng/wp-content/uploads/IssawiyaReport.pdf. Accessed August 2013 (Hebrew).

Bollens, S. A. (1998) Urban Planning amidst ethnic conflict: Jerusalem and Johannesburg. *Urban Studies*, 35(4): 729–50.

Bollens, S. A. (2000) On narrow ground: urban policy and ethnic conflict in Jerusalem and Belfast. Albany: State University of New York Press.

Bollens, S. A. (2007) *Comparative Research on Contested Cities: Lenses and Scaffoldings*. London: LSE Crisis States Research Centre. Available from: www.lse.ac.uk/international Development/research/crisisStates/download/wp/wpSeries2/wp172.pdf. Accessed May 2014.

Bollens, S. A. (2012) *City and Soul in Divided Societies*. London and New York: Routledge.

Botkyrka Statistics (2014) *Area Statistics Fittja*. Available from: http://www.botkyrka.se/kommunochpolitik/ombotkyrka/kommunfakta/botkyrkaisiffrorstatistik. Accessed August 2014.

Braverman, I. (2006) Illegality in East Jerusalem: between house demolitions and resistance. *Theory and Criticism*, 28: 11–42 (Hebrew).

Calame, J., and Charlesworth, E. (2009) *Divided Cities: Belfast, Beirut, Jerusalem, Mostar, and Nicosia*. Philadelphia: University of Pennsylvania Press.

Castells, M. (1983) *The City and the Grassroots: A Cross-Cultural Theory of Urban Social Movements*. Berkeley: University of California Press.

Cohen Blankshtain, G., Ron, A., and Gadot Perez, A. (2013) When an NGO takes on public participation: preparing a plan for a neighborhood in East Jerusalem. *International Journal of Urban and Regional Research*, 37(1): 61–77.

Dalby, S. (1990) American security discourses: the persistence of geopolitics. *Political Geography Quarterly*, 9: 171–88.

Detaljplaneprogram för Fittja (2012) [Fittja Local Master Plan 2012] Botkyrka Municipality (Swedish).

Dumper, M. (2014) *Jerusalem Unbound: Geography, History and the Future of the Holy City*. New York: Colombia University Press.

Flyvbjerg, B. (2006) Five misunderstandings about case-study research. *Qualitative Inquiry*, 12(2): 219–45.

Flyvbjerg, B. (2011) The case study. In N. Denzin and Y. Lincoln (eds), *The Sage Handbook of Qualitative Research*, 4th edn. Thousand Oaks, CA: Sage: 301–16.

Forum for Cities in Transition (n.d.) Available from: http://citiesintransition.net/. Accessed October 2013.

Fregonese, S. (2009) The urbicide of Beirut? Geopolitics and the built environment in the Lebanese civil war (1975–1976). *Political Geography*, 28(5): 309–18.

Fregonese, S. (2012) Urban geopolitics 8 years on: hybrid sovereignties, the everyday, and geographies of peace. *Geography Compass*, 6(5): 290–303.

Gaffikin, F., and Morrisey, M. (2011) *Planning in Divided Cities*. London: Wiley-Blackwell.

Gaffikin, F., and Perry, D. (2012) The contemporary urban condition understanding the globalizing city as informal, contested, and anchored. *Urban Affairs Review*, 48(5): 701–30.

Gough, K. V. (2012) Reflections on conducting urban comparisons. *Urban Geography*, 33(6): 866–78.

Graham, S. (ed.) (2004) *Cities, Wars and Terrorism, Towards an Urban Geopolitics*. New Jersey: Blackwell.

Haaretz (2013) *Jerusalem Digs Ditch to Separate Jewish and Palestinian Neighborhoods* (Nir Hasson, Haaretz.co.li, 13 September). Available from: www.haaretz.com/news/national/.premium-1.546801. Accessed October 2013.

Hall, T., and Vidén, S. (2005) The Million Homes Programme: a review of the great Swedish planning project. *Planning Perspectives*, 20(3): 301–28.

Harsman, B. (2006) Ethnic diversity and spatial segregation in the Stockholm region. *Urban Studies*, 43(8): 1341–64.

Hasson, S. (ed.) (2007) *Jerusalem in the Future: The Challenge of Transition*. Jerusalem: Floersheimer Institute for Policy Studies.

Hepburn, A. C. (2004) *Contested Cities in the Modern West*. New York: Palgrave.

ICBS (2014) *Yearly Report 2014*. Available from: www.cbs.gov.il. Accessed September 2013 (Hebrew).

Jerusalem Municipality (2013) *Jerusalem Municipality Website*. www.jerusalem.muni.il. Accessed September 2013.

Jerusalem Municipality Planning Department (2004) *Jerusalem Master Plan 2000*. Jerusalem: Jerusalem Municipality.

Khamaisi, R. (2002) Shared space, separate geo-politically: Al-Quds Jerusalem capital for two states. *Geoforum*, 33(3): 278–83.

Khamaisi, R. (2010) Sustainable in Jerusalem, International Society of City and Regional Planners. *The 46th ISOCARP Congress 'Sustainable City Developing World'*. Nairobi, Kenya, 19–23 September.

Lapidoth, R. (2006) Jerusalem. *Max Planck Encyclopedia of Public International Law*. Oxford: Oxford University Press. Available from: www.mpepil.com. Accessed September 2013.

Legeby, A. (2013) Patterns of co-presence: spatial configuration and social segregation. PhD diss. KTH School of Architecture, Stockholm: KTH.

Lloyd, C., Shuttleworth, I., and Won, D. (eds) (2014) *Social-Spatial Segregation*. London: Policy Press.

Lundström, M. J., Fredriksson, C. and Witzell, J. (eds) (2013) *Planning and Sustainable Urban Development in Sweden*. Stockholm: Swedish Society for Town and Country Planning.

McFarlane, C., and Robinson, J. (2012) Introduction: experiments in comparative urbanism. *Urban Geography*, 33(6): 765–73.

Marcuse, P. (1995) Not chaos but walls. In S. Watson and K. Gibson (eds), *Post Modern Cities and Spaces*. Cambridge: Blackwell: 243–54.

Marcuse, P. (2002) The partitioned city in history. In P. Marcuse and R. van Kempen (eds), *Of States and Cities: The Partitioning of Urban Space*. Oxford and New York: Oxford University Press: 11–34.

Marcuse, P. (2006) Enclaves yes ghettos no: segregation and the state. In D. Varady (ed.), *Desegregating the City: Ghettos, Enclaves and Inequality*. Albany: State University of New York Press: 15–30.

Marcuse, P., and van Kempen, R. (eds) (2002) *Of States and Cities: The Partitioning of Urban Space*. Oxford and New York: Oxford University Press.

Margalit, M. (2006) *Discrimination in the Heart of the Holy City*. Jerusalem: International Peace and Cooperation Centre.

Musterd, S. (2005) Social and ethnic segregation in Europe: levels, causes, and effects. *Journal of Urban Affairs*, 27(3): 331–48.

Musterd, S., and Ostendorf, W. (eds) (2013) *Urban Segregation and the Welfare State: Inequality and Exclusion in Western Cities*. London: Routledge.

Nightingale, C. H. (2012) *Segregation: A Global History of Divided Cities*. Chicago: University of Chicago Press.

Nijman, J. (2007) Introduction: comparative urbanism. *Urban Geography*, 28(1): 1–6.

Nijman, J. (2015) The theoretical imperative of comparative urbanism: a commentary on 'Cities beyond Compare?' by Jamie Peck. *Regional Studies*, 49(1): 183–6.

Nolte, A., and Yacobi, H. (2015) Politics, infrastructure and representation: the case of Jerusalem's light rail. *Cities*, 43: 28–36.

Parnell, S., and Robinson, J. (2012) (Re)theorizing cities from the Global South: looking beyond neoliberalism. *Urban Geography*, 33(4): 593–617.

Peck, J. (2015) Cities beyond compare? *Regional Studies*, 49(1): 183–6.

Petersson, L. (2011) *More Crime in LUA Areas*. Välfärd, 3. Available from: www.scb.se/valfard. Accessed August 2013 (Swedish).

Pickvance, C. G. (1986) Comparative urban analysis and assumptions about causality. *International Journal of Urban and Regional Research*, 10(2): 162–84.

Pickvance, C. G. (2001) Four varieties of comparative analysis. *Journal of Housing and the Built Environment*, 16(1): 7–28.

Pierre, J. (2005) Comparative urban governance: uncovering complex causalities. *Urban Affairs Review*, 40(4): 446–62.

Pintamo, H. (2011) *Background Material for Fittja Public Consultation*. Botkyrka: Botkyrka Municipality (Swedish).

Planning Administration (2013) *Israeli Planning Administration Annual Report 2012*. Jerusalem: Ministry of the Interior. Available from: http://www.moin.gov.il/SubjectDocuments/shnaton_203.pdf. Accessed August 2013.

Porter, L. (2014) Possessory politics and the conceit of procedure: exposing the costs of rights under conditions of dispossession. *Planning Theory*, 13(4): 387–406.

Pullan, W., and Baillie, B. (eds) (2013) *Locating Urban Conflicts: Ethnicity, Nationalism and the Everyday*. London: Palgrave Macmillan.

Ramberg, I., and Pripp, O. (2002) *Fittja: The World and the Everyday*. Botkyrka: Botkyrka Multicultural Centre (Swedish).

Robinson, J. (2002) Global and world cities: a view from off the map. *International Journal of Urban and Regional Research*, 26(3): 531–54.

Robinson, J. (2006) *Ordinary Cities: Between Modernity and Development*. London: Routledge.

Robinson, J. (2011) Cities in a world of cities: the comparative gesture. *International Journal of Urban and Regional Research*, 35(1): 1–23.

Robinson, J. (2014) Introduction to a virtual issue on comparative urbanism. *International Journal of Urban and Regional Research*. Available from: http://onlinelibrary.wiley. com/doi/10.1111/1468-2427.12171/abstract. Accessed August 2015.

Rokem, J. (2011) The *Jerusalem Master Plan 2000*: future challenges and opportunities. *Il Giornale dell'Architettura*, 10(99) (Italian).

Rokem, J. (2013) Politics and conflict in a contested city: urban planning in Jerusalem under Israeli rule. *Bulletin du Centre de recherche français à Jérusalem*, 23.

Rokem, J. (2016) Beyond incommensurability: Jerusalem and Stockholm from an ordinary cities perspective. *CITY*, 20(3): 451–61.

Rokem, J., and Allegra, M. (2016) Planning in turbulent times: exploring planners agency in Jerusalem. *International Journal of Urban and Regional Research*. doi:10.1111/1468-2427.12379.

Rokem, J. and Vaughan, L. (2017) Segregation, mobility and encounters in Jerusalem: the role of public transport infrastructure in connecting the 'Divided City', *Urban Studies*. doi: https://doi.org/10.1177/0042098017691465.

Roy, A. (2009) The 21st-century metropolis: new geographies of theory. *Regional Studies*, 43(6): 819–30.

Roy, A. (2014) *Before Theory: In Memory of Janet Abu-Lughod*. Available from: www. jadaliyya.com/pages/index/16265/before-theory_in-memory-of-janet-abu-lughod. Accessed February 2014.

Sampson, Robert, J. (2013). *Great American City: Chicago and the Enduring Neighborhood Effect*. Chicago: University of Chicago Press.

SCBS (Swedish Central Bureau of Statistics) (2014) Statistiska Sentralbyro *Yearly Report 2014* Statistiska Sentralbyro. Available from: www.scb.se. Accessed September 2014 (Swedish).

Scott, A. J., and Stroper, M. (2014) The nature of cities: the scope and limits of urban theory. *International Journal of Urban and Regional Research*, 39(1):1–16.

Sheppard, E., Leitner, H., and Maringanti, A. (2013). Provincializing global urbanism: a manifesto. *Urban Geography*, 34(7): 893–900.

Shlay, B. A., and Rosen, G. (2015) *Jerusalem: The Spatial Politics of a Divided Metropolis*. Cambridge: Polity Press.

Sidaway, J. D. (2009) Shadows on the path: negotiating geopolitics on an urban section of Britain's South West Coast Path. *Environment and Planning D*, 27(6): 1091–116.

Stockholm Office of Research and Statistics (2015) *Stockholm Statistical Yearbook (2014)*. Stockholm: Klippan Press. Available from: www.statistikomstockholm.se/attachments/ article/38/arsbok_2012.pdf. Accessed August 2013 (Swedish).

Storstadskommittén (1997) *Delade Städer SOU 1997:118* [Divided Cities Report]. Available from: www.riksdagen.se/sv/Dokument-Lagar/Utredningar/Statens-offentliga-utredningar/sou-1997-118-_GLB3118/. Accessed August 2013 (Swedish).

UNCTAD (United Nations Conference on Trade and Development) (2013) *The Palestinian Economy in East Jerusalem* [PDF]. New York and Geneva: United Nations. Available from: http://unctad.org/en/PublicationsLibrary/gdsapp2012d1_en.pdf. Accessed August 2013.

van Kempen, R. (2007) Divided cities in the 21st century: challenging the importance of globalization. *Journal of Housing and Built Environment*, 22: 13–31.

van Kempen, R., and Murie, A. (2009) The new divided city: changing patterns in European cities. *Tijdschrift voor economische en sociale geografie*, 100(4): 376–576.

Vaughan, L., and Arbaci, S. (2011) The challenges of understanding urban segregation. *Built Environment*, 37(2): 128–38.

Wacquant, L. (1997) Three pernicious premises in the study of the American ghetto. *International Journal of Urban and Regional Research*, 21(2): 341–53.

Wacquant, L. (2008) *Urban Outcasts: A Comparative Sociology of Advanced Marginality*. Cambridge: Polity Press.

Wacquant, L. (2014) Marginality, ethnicity and penality in the neoliberal city: an analytic cartography. *Ethnic and Racial Studies*, 37(10): 1687–711.

Ward, K. (2008) Editorial: toward a comparative (re)turn in urban studies? Some reflections. *Urban Geography*, 29(5): 405–10.

Ward, K. (2010) Towards a relational comparative approach to the study of cities. *Progress in Human Geography*, 34(4): 471–87.

Watson, V. (2013) Planning and the 'stubborn realities' of global south-east cities: some emerging ideas. *Planning Theory*, 12(1): 81–100.

Watson, V. (2014) The case for a southern perspective in planning theory. *International Journal of E-Planning Research*, 3(1): 23–37.

Yacobi, H. (2009) Towards urban geopolitics. *Geopolitics*, (14): 576–81.

Yacobi, H. (2012) Borders, boundaries and frontiers: notes on Jerusalem's present geopolitics. *Eurasia Border Review*, 3(2): 55–69.

Yacobi, H., and Pullan, W. (2014) The geopolitics of neighbourhood: Jerusalem's colonial space revisited. *Geopolitics*, 19(3): 514–39.

Yin R. K. (2009) *Case Study Research: Design and Methods*. London: Sage.

Part II

Urban geopolitics
South and South East Asia

Camillo Boano

This section continues the reflections on comparativism, offering the reader another set of 'differences' from whence to learn about positioning the variegated contexts of southern Asia as "new urban epicentre" (Robinson, 2011: 595), reflecting on multidirectional learning across contexts where both the exploration of contingent and universal realities are taking place simultaneously, from the specific, grounded and situated spatial practices and realities to wider conversations on contested urban life. Grounded in East Asian and South Asian spatial and urban realities, the three chapters in this section grasp the complexities of multicultural developments and their contested natures. The first chapter, Chapter 4, written by Sadaf Sultan Khan, Kayvan Karimi and Laura Vaughan, discusses their detailed spatial investigation of Karachi, the capital of Pakistan, where language, ethnicity, spatial clustering and politics are intrinsically linked, dissecting the impact a national minority has had enabling a transformation of the city's geopolitics and its urban forms.

Moving to the next chapter, Chapter 5 is an in-depth investigation by Pawda F. Tjoa of contested urban space through the lens of 'marketplace coordination' in Jakarta, highlighting the roots of social tension and the escalation of internal conflict during the period 1997–8. It sheds light on the ideology of 'development' as a catalyst for conflict, which contributed to the persistence of urban geopolitical fragmentation.

The last chapter in this section, Chapter 6, by Apurba Kumar Podder explores a case of an illegal bazaar located in Khulna, one of the southern cities in Bangladesh. It offers a different angle – the radical rethinking of one aspect of poverty culture: 'doing nothing'. In a local ethnographic exploration, it is suggested doing nothing should be understood as an alternative mode of the poor's occupational urban geopolitics to sustain in a condition of unequal power relation in urban areas.

References

Robinson, J. (2011) Cities in a world of cities: the comparative gesture. *International Journal of Urban and Regional Research*, 35(1): 1–23.

4 The tale of ethno-political and spatial claims in a contested city

The *Muhajir* community in Karachi

Sadaf Sultan Khan, Kayvan Karimi and Laura Vaughan

Introduction

Karachi, today, is essentially a city of migrants, the result of successive waves of in-migration triggered by past events and decisions that took place predominantly on the national and the international political stage. Possibly the most significant of these was the departure of the British Raj and the Partition of the Indian sub-continent in 1947 to the separate states of India and Pakistan. This resulted in a large influx of *Hindustani* migrants into the city where over a million people arrived as it became the federal capital of the new nation of Pakistan. The *Muhajir* or refugees – primarily Urdu-speaking, north Indian Muslims migrating from Hindu-majority urban centres in previously undivided India – inundated the city, transforming it from a small, cosmopolitan colonial port to a city that appeared to have morphed into the various neighbourhoods of Lucknow, Agra and Hyderabad from where these refugees had migrated (Siddiqi, 2008). Today, their descendants continue to be referred to as *Muhajir*.

While the *Muhajir* community is the largest migrant group in Karachi – they have consistently made up over 50 per cent of the city's population since Pakistan's first census in 1951 – today they represent only 6 per cent of Pakistan's total population. Since arrival post-Partition, the *Muhajir* have continued to be viewed as a landless, rootless people whose loyalties to Pakistan are suspect. This is due to a combination of their ongoing geographical concentration and their signature urban Indian heritage, which is in marked contrast to the clan and tribal solidarities of many of Karachi's domestic migrant communities who still exhibit strong ties to ancestral lands. This imposed and perceived *otherness* has resulted in the *Muhajir* community behaving as a people under constant threat, manifesting in the form of persistent geographical concentration over time, and political mobilization to demand and preserve their rights.

It could be said that Karachi's population is the outcome of the geopolitics of the region: Partition-related refugee communities; rural domestic migrants displaced by green revolution policies in the 1960s; Biharis, Bengalis and Afghans displaced by war, secession and occupation in Bangladesh and Afghanistan in the 1970s and 80s; and, most recently, internally displaced people (IDPs), the result of military operations in Pakistan's northern areas. This has brought a diverse array of communities and cultures into close contact in an urban environment where resources such as housing, transport and employment are in short supply. Non-state actors, through a process of politicization of ethno-linguistic solidarities, have stepped in to bridge the disparity between demand and supply. This complex situation suggests that migrant politics may be critical to the shaping of the city where language, ethnicity, spatial clustering and politics are intrinsically linked; language often influences ethnic loyalties, which, in turn, sway political affiliation. In essence, through an analysis of the spatial and ethno-political histories of the *Muhajir* community in Karachi, this chapter seeks to dissect the impact the city's often volatile geopolitics has had on the form it takes today.

The *Muhajir* community's process of migration and settlement seemingly follows the trajectory outlined by Vertovec (1995): (1) migration, (2) the establishment of cultural institutions, (3) political mobilization and, finally, (4) identity decline and rejuvenation. In the absence of detailed ethnographic data, where the last useable census was conducted in 1998 and published data are available at the very large scale of the city district,[1] analysis of the processes of migration and settlement as outlined by Vertovec is central to studying the settlement patterns of the *Muhajir* community in the city today. This was done through the mapping and analysis of the clustering of certain cultural/communal institutions particular to the *Muhajir* community.

Minority or ethnic clustering is a complex spatial phenomenon, often a combination of both forced separation by the majority group of peoples perceived as the 'other' to maintain a sense of purity of the host community (Sibley, 1995; Sennett, 1996) and a self-imposed separation by the minority for the purposes of socio-cultural preservation (Werbner, 2005). The location, persistence and densification of the ethnic settlement may be shaped by market forces pertaining to affordable housing (Phillips, 2006), proximity to employment (Charalambous and Hadjichristos, 2011), security and safety in numbers, resulting in an inversion of power within the enclave in favour of the minority (Peach, 1996), or all of the above. Affordable commutes and access to reliable employment, services and social networks are features that are built over long-term residence and are advantages that diminish dramatically, should the residents sell out and move away (Simone, 2013).

Such clustering may be beneficial for the purposes of acclimatization of fresh arrivals and for the preservation and persistence of cultural traditions; depending on the location of the clusters, it may also lead to a reduction in interaction between the residents and the host community and more limited economic and social opportunities. This may be particularly problematic if immigrant enclaves are also spatially isolated (Legeby, 2009). Such a situation may heighten the sense of marginalization among members of the community, leading to the crystallization of political identities, consolidation of ethno-religious solidarities and the perpetuation of feuds and prejudices (Shirlow and Murtagh, 2006).

This chapter presents an overview of urban geopolitical and developmental histories of post-Partition Karachi in order to show how in-migration of multiple communities and the ethno-political affiliations of various state-backed actors have impacted the planning and development of the city. This process has, over time, laid the groundwork for the city's current socio-spatial divisions. Additionally, the connection between language, ethnicity and politics and how this contributes to ethno-spatial appropriation and contestation, ethnicity-based service monopolies and uneven planning and development of the city will also be discussed. By viewing the urban geopolitics of the city through the spatial-political trajectory of the *Muhajir* community in Karachi since Partition, this chapter examines the synergistic and often divisive relationship between ethnicity, politics and urban development, an issue shared by numerous contested, post-colonial urban environments today.

Historical background

As suggested by the title, this chapter describes a study into the nature of the relationship between ethno-political identities and the occupation and adaptation of space in Karachi. In order to deconstruct this relationship, there is a need to establish that there is, in fact, such connection – and, specifically, between the emergence of urban geopolitical ethnic identities and activities and the spatial planning and development of the city. Table 4.1 outlines broad political periods in Pakistan's history, as well as presenting in-migration, urban development and the ethno-political outcomes in the city during these periods in order to illustrate how these issues are interconnected and how the politics of ethnicity have coloured official patronage and development of the city and its communities.

Table 4.1 shows that, with each wave of in-migration, the state proposed a master plan to address growing concerns of housing, transport and infrastructure development. At each point these interventions fall short of the needs of a growing population or are abandoned, with the informal sector stepping in to bridge the gap. It also illustrates how the movement of migrants into the city was intrinsically linked to the politics of post-colonial development. Additionally, it appears that state patronage shifted from community to community, depending upon the affiliations of those in power. Events like Pakhtun army man Ayub Khan awarding transport licences to members of his community in the 1960s, resulting in a long-term transport monopoly by the Pakhtun community, illustrate this. Similarly, in the 1970s Sindhi feudal Bhutto promotes Sindhi language and Sindhi presence in public sector positions and, finally, *Muhajir* military man Pervez Musharraf backs the MQM and pours funds into infrastructure development in *Muhajir*-dominated Karachi in the early 2000s. This state-led patronage has assisted in the emergence of ethnic trade and service monopolies and the subsequent politicization of ethnic identities.

Karachi's unholy trinity: language, ethnicity and politics

Karachi's post-Partition history is characterized by almost constant in-migration, developing ethnic trade and service monopolies and political marginalization.

Table 4.1 Synthesis of Karachi's political and urban development histories

Year	Political	Migration outcomes	Urban development plans	Ethno-political outcomes
1947–57	7 prime ministers in 11 years. **Ethnicity:** Primarily migrant backgrounds. Iskander Mirza declares Martial Law in October 1958	**Arrivals:** *Muhajir* **Status:** Refugee **Cause:** Partition of the Indian sub-continent, 1947	**Master plan:** MRV Plan 1951 (not implemented) Refugee rehabilitation is the primary objective of the government Settlements can be broadly classified as (1) Relief and transit camps, (2) government-sponsored housing schemes, (3) community-initiated housing societies and (4) informal squatter settlements	*Muhajireen* were clustering in northern and north-eastern areas of the city, e.g. Liaquatabad, Nazimabad, North Nazimabad, Federal B Area, as well as placed-based housing societies of Pakistan Employees Cooperative Housing Society (PECHS), Sindhi Muslim, Bahadurabad etc. at the edge of the city centre and the squatter settlements on the northern banks of the Lyari River – Golimar, Lalukhet etc.
1958–69	**President:** Field Marshal Ayub Khan **Affiliation:** Army **Ethnicity:** Pakhtun **Elections:** 1965	**Arrivals:** Pakhtun/ Punjabi **Status:** Economic migrant **Cause:** Green revolution policies implemented in rural areas. Construction of Mangla and Tarbela Dams. Industrialization of urban centres	**Master plan:** Greater Karachi Resettlement Plan (GKRP) (1958) Karachi Master Plan 1974–85 (1964) Government decides to 'decentralize population' and hence Doxiades GKRP proposes industrial estates at the periphery (Landhi-Korangi and New Karachi), with associated residential units to house displaced inner city squatters Karachi Master Plan 1974–85 proposes insightful transport and infrastructure interventions, but is shelved shortly after initiation	Government employs a policy of spatial and commercial marginalization of the *Muhajir* community by (1) decanting inner city settlements to industrial estates at the periphery, (2) moving the federal capital from Karachi up-country to Islamabad, (3) preferential treatment of Pakhtuns in the awarding of transport licences Emergence of ethnic tensions as *Muhajirs*, Pakhtuns and Punjabis now vie for the same resources and economic opportunities in the city

		Arrivals	Master plan	
1970–7	**PM:** Z. A. Bhutto **Affiliation:** PPP **Ethnicity:** Sindhi **Elections:** 1970, 1977	**Arrivals:** Bihari/Bengalis **Status:** Refugee **Cause:** Secession of East Pakistan and the creation of Bangladesh	**Master plan:** – Coastal development projects initiated, along with changes in floor area ratios (FARs) for certain areas to boost commercial activity, result in the emergence of the high-rise apartment buildings. Construction of Metroville projects as low-cost housing and the subsequent mushrooming of informal settlements in proximity to these government-sponsored projects to the west of the city	Further marginalization of the *Muhajir* community through the implementation of policies advocating (1) Sindhi as an official language in schools and government departments, thus contesting the status of Urdu as the national language, (2) the re-evaluation of the quota system to give rural communities more access to government jobs and university places, thereby reducing quotas for urban communities such as the *Muhajireen*, (3) the nationalization of industries and businesses, resulting in increased competition for positions, in keeping with the quota system This resulted in the language riots of 1972 and the emergence of student politics in government educational institutions
1977–88	**President:** General Zia-ul-Haq/Martial Law **Affiliation:** Army **Ethnicity:** Punjabi **Elections:** 1985 (non-partisan)	**Arrivals:** Afghans **Status:** Refugee **Cause:** Soviet occupation of Afghanistan	**Master plan:** Regularization of informal settlements (1978). The development of high-rise apartment complexes by private developers in Gulshan and Gulistan-e-Jauhar areas in the north east of the city, financed by ex-pats' remittances from the Gulf States	*Muhajir* students leave the Punjabi-centric Islami Jamiat-e-Talaba (IJT) and form the All Pakistan Muttahida Students Organization (APMSO) in 1978 at Karachi University. Establishment of the Muttahida Qaumi Movement (MQM) (1984), which goes on to sweep the first elections they contest in 1988 in Karachi and Hyderabad. Ethnic tensions rise, resulting in clashes between *Muhajirs* and Pakhtuns, especially in the squatter settlements of Qasba and Orangi to the west of the city, between 1985 and 1986.

(Continued)

Table 4.1 Synthesis of Karachi's political and urban development histories (*Continued*)

Year	Political	Migration outcomes	Urban development plans	Ethno-political outcomes
1988–99	**PM:** Benazir Bhutto, Nawaz Sharif **Affiliation:** PPP/PML **Ethnicity:** Sindhi/Punjabi **Elections:** 1988, 1990, 1993, 1997	**Arrivals:** –	**Master plan:** Karachi Development Plan 2000 (1989) Development of Gulshan-e-Iqbal, Scheme 45, Taiser Town and Gulzar-e-Hijri again in the north-eastern quadrant of the city	MQM's Dr Farooq Sattar is mayor of Karachi from 1987–92. A paramilitary operation is conducted to 'cleanse Karachi of anti-social elements', targeting primarily *Muhajir* majority areas, known as Operation Clean-up (1992–4), resulting in thousands of arrests and extra-judicial killings of MQM activists. The party's top brass went 'underground' and party chairman, Altaf Hussain, went into self-imposed exile. During this period areas like Lines Area, Liaquatabad and Golimar become notoriously known as 'no-go' areas due to clashes between *Muhajir* youths and the city's security forces
1997–2007	**President:** General Pervez Musharraf **Affiliation:** Army **Ethnicity:** Muhajir **Elections:** 2002	**Arrivals:** Pakhtun **Status:** IDPs **Cause:** Anti-terrorism military operations in Swat and Waziristan	**Master plan:** Master Plan 2020 (2007) Flyovers and underpasses, the LEW and Northern bypasses are built, Dolmen City is revived and Creek City initiated Establishment of new displaced persons townships in Malir and Hawkesbay Official commercialization of 17 major commercial streets across the city	With MQM's Syed Mustafa Kamal as the city's mayor from 2005–10, MQM dominates the City District Government and is seen as the king-maker of the nation. Large-scale infrastructure projects that had seen little movement are brought to life and completed

Sources: Karachi Master Plan 2000, Ansari (2005), Gayer (2014), Hasan (1999), Zamindar (2010).

This section discusses the critical relationship between language, ethnicity and politics and its role in contestation, spatial demarcation and socio-political dominance in Karachi today, using the *Muhajir* community as a point of reference in the emergence and evolution of ethno-political identities in the city.

The *Muhajir* community today can broadly be defined as first-wave Muslim migrants and their descendants, who originated primarily from urban centres in India's Uttar Pradesh, Central Provinces, Hyderabad Deccan, Rajasthan and Gujarat – essentially, Hindu-majority provinces at the time of Partition. Hence, this is a community that is comprised of a diverse array of ethnic sub-groups. Their commonality at the time of their migration and settlement in Karachi was a shared Islamic faith; Urdu as the *lingua franca* of the Indian Muslim community; and a political conviction in the Pakistan Movement, the political movement spearheaded by the All-India Muslim League to demand a homeland for Indian Muslims once the British withdrew from India. Hence, the community's inception and cohesion was based on broad religio-political and linguistic similarities.

In the early stages of settlement in Pakistan, the label *Muhajir* was used as an official description of a 'situation' rather than as a reference to the identity or place of origin of these refugee communities. The *Muhajir* community's post-Partition access to political power, public sector institutions and economic opportunities was increasingly curtailed by various local power players, coupled with the arrival of newer migrant groups. This engendered a growing sense of both physical and political marginalization, resulting in the community resorting to the form of ethno-political identification common in Karachi, namely the emergence of the Muttahida Quami Movement (MQM) in 1984. Gone was the image of the landless, bedraggled refugee community to be replaced by a politically active, cultured, Urdu-speaking urban middle class, Pakistan's so-called fifth ethnicity (Ahmed, 1998). This was now a community demanding political recognition alongside Pakistan's four acknowledged ethnic groups – Balochis, Pakhtuns, Punjabis and Sindhis – each identified by a distinct language(s), culture and political representation. Hence, their reimagining as a cultured, Urdu-speaking, urban middle-class political entity has been critical to describing and mapping Karachi's ethno-political fault lines.

Associated with these social characteristics of the *Muhajir* community are certain tangible, spatial features such as the political party office and the religious building. These communal institutions not only provide a means of community building and identification, but can also act as socio-spatial proxies indicative of a community's presence and activities within an urban environment (Waterman and Kosmin, 1988).

Language

Urdu is a language with a significant political history, often considered to be a pidgin or Creole (i.e. mixed in its origin) language developed in the military encampments of the Mughal army, enabling soldiers of diverse ethnic backgrounds – Turks, Persians, Arabs, Indians etc. – to communicate. Its vocabulary borrowed from local languages such as Hindvi/Dehlavi, as well as from Arabic, Persian and Turkish. Contrary to this, Rahman (2011) argues that while Urdu and Hindi share the same Hindvi/Dehlvi roots, it was Mughal patronage of the language after the British replaced Persian with vernacular languages for official purposes that purged it of its Sanskrit vocabulary and pushed it towards the 'Arabicized/Persianized' form it takes today, both in its script and the hybridity of its vocabulary, so that it has become a symbol of the Muslim elite. With the politicization of Muslim identity in the early twentieth century culminating in the Pakistan Movement, Urdu became more closely linked to Indian Muslim nationalism through its use in printed religious and political literature (ibid.).

Post-Partition, language has become a politically contentious issue in Pakistan; Urdu was installed as the national language across both West and East Pakistan (subsequently Bangladesh) despite the numeric dominance of Bengali-speakers, resulting in recurring protests in Dhaka from 1952 until Bengali was given official language status in East Pakistan in 1956. Subsequent language-related policies include the compulsory teaching of Sindhi in schools and colleges and its recognition as an official language on a par with Urdu in Sindh in 1972, Sindh being the only province where Urdu speakers were in the majority. These state-sponsored initiatives resulted in the language riots where Urdu-speaking communities, encouraged by the Urdu press, protested. The *Muhajir* community consider their Urdu-speaking heritage as a badge of honour, indicative of their commitment to and sacrifice for Pakistan, while the native communities see it as a mark of *Muhajir* foreignness. Thus, this linguistic difference has become a means of differentiating between the sons of the soil (Ansari, 1995) and new migrants. This clash of cultures has only increased with time and the role of language in this discourse has only become more contentious in the way in which it provides a form of self-identity as well as a way to establish status, as has been the case with other migrant populations elsewhere in the world (Kershen, 2005).

The fact that Urdu remains the language of the *Muhajireen* to the present day is illustrated by census statistics (1998), where only 7.57 per cent of the total population claimed Urdu as their mother tongue in contrast to 48.52 per cent of Karachi's population claiming the same. Further analysis of census data shows that Urdu-speaking communities appear to be concentrated in two of Karachi's five districts – districts Central and East (see Figure 4.1a). This is one of the first indications for the continuing spatial clustering of the community within Karachi.

Institutional politics

As the discussion thus far has shown, in Pakistan, language and ethnicity are intimately connected, and this link has been exploited for political purpose, as Table 4.1 shows, through the implementation of state-sponsored policies. With regard to party politics in Karachi, ethno-linguistic solidarities are further exploited. Two broad categories of political parties exist: mass-based parties – both religious and secular – and ethnicity-based parties. In Karachi, even mass-based secular parties like the Pakistan People's Party (PPP) exhibit an ethnic bias, hence, until recently, Sindhis and Balochis voted primarily for the PPP. Among many other such ethnic affiliations, *Muhajir* communities have historically voted for the MQM since the party's foundation (Hasan, 2005). In recent times, Karachi has seen the aggressive re-emergence of various right-wing Sunni groups like the Pakistan Sunni Tehreek (PST) and the Jamaat-e-Islami (JI), where the PST is looking to break into the *Muhajir* vote base while the JI are in the midst of a violent attempt to reclaim the *Muhajir* vote bank they lost in the 1980s.

By reviewing electoral results from 1988–2008, a consistent pattern emerges that shows a concurrence between linguistic concentrations and voting patterns; the city's Central and Eastern districts[2] not only show a majority of Urdu-speaking households, but also consistently vote overwhelming for the MQM (see Figure 4.1b). This seems to reinforce the notion that, in Karachi, there has been a trans-formation of ethno-linguistic solidarities into urban geopolitical ethnic identities.

The MQM's physical political presence in many of the city's areas can be mapped too due to the party's unique three-tier structure. This consists first of Nine Zero, their nationwide headquarters located in Azizabad, a middle-income neighbourhood in Karachi's District Central. Under the headquarters are 26 city-wide sector offices, each overseeing eight to ten unit offices embedded within the neighbourhood they serve. The bottom tier, comprising unit offices, is of particu-lar interest to this study; their location within neighbourhood streets is indicative of political activity at that scale, where political activists are youths from the area who therefore possess a familiarity and access to both space and information. The locations of these units within the city showed that the highest concentrations existed in Karachi's districts Central and East, exhibiting similar patterns to those shown earlier through linguistic mapping and electoral results (Figure 4.1c).

Religion

Language and political affiliation provide the broad strokes of community defini-tion in Karachi, as can be seen from the number of religious groups with a political presence in the previous section. Religion too plays a significant role in the defini-tion of a community. In the context of the *Muhajir*, among the various sects within the Islamic faith, two are of particular interest: the Shi'a and Barelvi communities. The former is the second largest denomination in Islam and is mainly comprised of the followers of Ali ibn Abi Talib, Prophet Mohammed's son-in-law and the fourth Caliph, and the latter community practices a form of Sunni Islam that was founded in the north Indian city of Bareilly by Ahmed Raza Khan. It is important to note that while the *Muhajireen* are not the only practitioners of these schools of belief in Karachi, prior to Partition, both sects had large followings in the regions from which the *Muhajireen* originated. Hence, in Karachi today, many of those who subscribe to either the Shi'a or Barelvi schools of religious thought are of *Muhajir* descent (Jones, 2007; Robinson, 2014; Verkaaik, 2004). In the Islamic faith, due to the frequency of congregational prayers, religious buildings are a common neighbourhood feature and practising Muslims generally reside close to a mosque catering to their particular denomination. Thus, the analysis of the distribution of these particular religious institutions provides another source of information on the spatial patterning of the *Muhajir* community.

Mapping religious institutions for both the Barelvi mosque and the Shi'a place of worship, the *imambargah*, showed that – as was the case with linguistic and political clustering – the highest concentrations of Barelvi mosques and *imambargah*s were to be found in both districts Central and East (see Figure 4.1d and e). This firmly corroborates the findings of the linguistic analysis, which suggests that the *Muhajireen* appear to be clustering in specific districts of the city despite the fact that the census data used is over 15 years out of date.

Establishment of a community: distribution and clustering of communal institutions

While the use of communal institutions and features such as language, MQM unit offices and Barelvi mosques and Shi'a *imambargah*s help to describe community concentrations at the district scale, these land parcels are large and municipal boundaries lack the nuances required to illustrate actual community clusters. Hence, in order to identify specific community clusters, geographical information system (GIS) analysis is used to statistically capture measurable clusters of features or activities that can be attributed to *Muhajir* population communal activities. It is important to point out that by using such communal institutions as socio-spatial proxies to locate the community in the city today, in the absence of reliable ethnographic data, it is also possible to illustrate the intimate connection between ethnicity and politics through their spatial overlaps.

A few things should be noted at this point. First, that the community continues to occupy the localities in which they originally settled shortly after arrival in the city, as outlined in the years 1947–58 in Table 4.1, while simultaneously colonizing newer adjoining localities. Second, many of the neighbourhoods exhibiting the highest concentrations of *Muhajir* communal institutions shown in Figure 4.2 were high-density, lower- to middle-income localities that repeatedly came up in news reports as centres of *Muhajir* agitation and major MQM activity, such as Liaquatabad, Golimar and Lines Area. Finally, there is a noticeable absence of *Muhajir* identity markers in the city's elite district to the south and the ethnically mixed informal settlements to the west. These features seem to suggest that, in addition to state-sponsored marginalization and inter-community contestation for public services, ethnic clustering, spatial proximity and economic characterization have been partially responsible for the emergence of a group identity, defined by politics, language and close proximity within itself. The latter point is explored further in the following section.

Reimagining a community: political power, location and urban control

This section focuses on the relationship between spatial configuration, ethnic clustering and the strategic positioning of a community within the fabric of the

Figure 4.1 District-wise distribution of (a) Urdu as a mother tongue, (b) election results 2008, (c) MQM unit offices, (d) *Imambargah*s and (e) Barelvi mosques

Sources: *Census Reports*, Government of Pakistan. Election commission reports accessed from ecp.gov.pk. Communal institutions mapped by the researcher for the years 2011–13

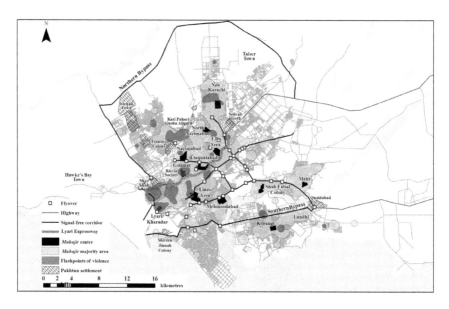

Figure 4.2 Location of the *Muhajir* majority areas, *Muhajir* centres, signal-free corridors
and flyover and underpass projects (Maher, 2013), as well as flashpoints of violent encounters between communities for the election period 1 May–31 July 2013

Source: Author, 2016

city. In 2002, a new devolution plan was instituted at the city district level, replacing Karachi's five-district structure with 18 towns under which there were 178
union councils, each union council comprised of 13 elected councillors. During
this period MQM had the largest number of elected councillors of any political
party in the City District Government, with this form of governance giving them
more access to political power and funds than they had experienced since 1992.
This resulted in a mushrooming of infrastructure development projects across
the city.

Location of major thoroughfares

The nature of transport infrastructure in changing the spatial landscape of Karachi
is an important element in tracing the spatial history of the *Muhajireen*. As part
of the Karachi Master Plan 1974–85, two transport links were first proposed in
1968. The intention was to enable the rapid movement of people and goods
between the port and Karachi's hinterland; the Southern and Northern Bypasses
(see Figure 4.2). This proposal, along with the master plan itself, was abandoned
shortly after its inception, only to be revisited in 2001 when both were built along
with the Lyari Expressway. The latter was initially proposed as an alternative to
the bypasses. The eventual development of these transport links was part of a

scheme to develop six signal-free corridors through Karachi. These proved to be highly controversial on several levels: first, by displacing a number of non-*Muhajir* communities living alongside the planned routes to more remote peripheral localities of the city (Taiser and Hawke's Bay towns); and, second, due to their consequential severance of previously connected central districts, resulting in a noticeably uneven development in the city. Figure 4.2 highlights, in particular, the flashpoints of violence that occur in close proximity to the expressway.

While these projects appear to be following the alignments of some of the city's busiest and longest thoroughfares connecting the centre and the port to the peripheries, they happen to be located primarily in the *Muhajir* majority areas found above, effectively connecting Karachi's peripheral *Muhajir* localities to its central stronghold. It should be noted that there has been limited intervention in the low-income, ethnically mixed settlements to the west of the city and the elite district to the south, with the bulk of the flyovers and underpasses required to facilitate these signal-free corridors being sanctioned and executed after 2005, when Karachi had a *Muhajir* mayor and MQM had a majority stake in the city government.

Location of major commercial centres

While initial *Muhajir* residential areas and commercial activities were confined at first to the old city centre, as the city has grown and population densities have shifted, newer commercial areas have emerged outside the city centre to cater to these changes. Unlike the commercial areas of the old city centre, many of the newer areas started as informal commercial areas. Through a long process that culminated in 2003, 17 streets city-wide were officially categorized as commercial: six major roads in phase 1 followed by 11 more in phase 2 (Anwar, 2010). Similarly to the flyovers, this policy was implemented while Karachi had a *Muhajir* mayor and an MQM majority in the city government. Again, as was the case with the major thoroughfares of the city, many of these streets are among the city's busiest thoroughfares, hence the logic of their selection for commercial activity seems obvious, yet the political advantage of the situation seems undeniable; of the 17 streets that were commercialized, only two are to be found outside the *Muhajir* majority areas (see Figure 4.3). These outliers are both situated in the city's elite localities of Defence and Clifton, while there is a noticeable absence of planned commercialization in the city's lower-income localities of Landhi and Korangi to the east and Orangi Town and Baldia to the west, the latter increasingly becoming areas of contention between the MQM and other ethnic and political groups active in these areas.

So who controls Karachi?

The analysis so far has highlighted two parallel processes: self-segregation by the *Muhajir* community and political and spatial marginalization by the state – for example, limiting access to public sector institutions by imposing a quota system

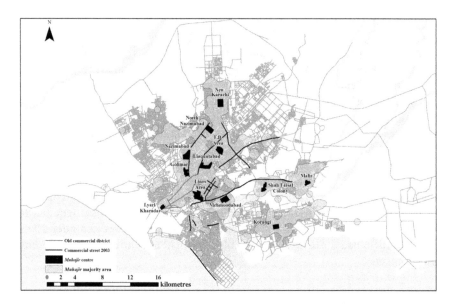

Figure 4.3 Streets commercialized in 2003, overlaid on *Muhajir* clusters R1,000m, showing that these areas fall primarily into *Muhajir*/MQM jurisdiction

Source: Author, 2016

that favoured a rural intake. Taken together, it seems evident that these processes
have contributed to the ongoing, spatial concentration of the community in areas
of the city, which has, in turn, helped sustain their spatial-political power in the
city over the last 30 years. But this does not mean that they are the only influential
ethno-political players present. The contestation of political power along the
city's myriad ethnic fault lines have led to conflict in focus points around the city,
making Karachi a dangerous megacity.

As the city has grown and newer migrant groups have moved in, large sections
of the peripheries have developed as informal settlements, housing new arrivals
where the state has failed to provide accommodation. Thus, many peripheral
settlements are multi-ethnic and low-income areas. Karachi is a dangerous city
where contestation of space and access to services has often resulted in violent
encounters between communities in the form of bombings, targeted shootings
and street clashes. Election time in the city is particularly volatile, with competing ethnic groups and political factions targeting their rivals to ensure votes.

For the purposes of this chapter, bombings for the election year (1 January–31
December 2013), as well as shootings for three months around the general election itself (1 May–31 July) were recorded and their locations plotted. A GIS clus-
ter analysis of these violent events, as seen in Figure 4.2, highlights particular
flashpoints in the city. These include certain peripheral *Muhajir* localities while
others are embedded deep within the *Muhajir* centre. On closer investigation of

those targeted in these encounters and the perpetrators, Karachi's ethno-political rivalries can be seen being played out through violent urban street battles. Encounters between *Muhajir* and non-*Muhajir* communities (Pakhtun, Baloch and Sindhi) in places like Kati Pahari and Qasba Colony in the north west and Shershah and Lyari in the east appear to be highlighted. Similarly, tit-for-tat shooting between rival *Muhajir* groups can be seen in New Karachi to the north, Lines Area in the centre and Korangi in the east. This seems to suggest that the MQM's political and physical strength wanes as distance from its political centre increases and that, while MQM may claim to be the dominant *Muhajir* representative, there is significant challenge to their dominance posed from within the community.

Conclusions

What appears to have happened in Karachi over the course of the last 65 years is a process of inversion of power, where a national minority has been able to control and transform the districts in which it constitutes a majority group. This local dominance has been achieved through the combined impact of legitimate political processes and violent street presence. This urban geopolitical response, in line with Vertovec's description of migrant resettlement patterns, has often been attributed to their systematic marginalization by the state. Nevertheless, it could be argued that this marginalization was not a passive process of minority exclusion by the state. Notably, the creation of MQM by a second-generation *Muhajir* community represents a capturing of power by a previously fervently nationalistic community, reimagining itself in its current ethno-political incarnation after being disenfranchised by the state.

The *Muhajir* community dominates the dense urban centre of the city, controlling transport links, key urban nodes and commercial and industrial sites. It is a political entity that is under threat at the peripheries of its domain, either from rival *Muhajir* political representatives such as the JI or Haqiqi Muthida Quamin Movement (MQM-H) or other ethnic groups eager for their piece of Karachi. This continuing contestation of power in the city can be said to show that political strength wanes as the distance from the political centre increases. Karachi's affluent residents in the southern-most areas of Defence and Clifton rarely vote consistently from one election to the next and these areas are, for the most part, devoid of any ethno-political violence.

Allegra *et al.* (2012) make a distinction between various types of so-called divided cities. They maintain that governance, decision-making and public policy are the points of distinction between city types, namely partitioned, contested and discrete cities. Whether it can be simply labelled as 'divided' is a moot point. Clearly, Karachi is a city in a state of constant turmoil, where the distribution and access to services and opportunities is determined by community numbers and ethno-political clout and where urban space is contested daily by individuals making decisions on residential location, as well as political powers determining urban policy. Despite repeated efforts by the state to limit their access to power and continuing challenges posed by other ethno-political factions, the *Muhajir* community continues to dominate Karachi both numerically and politically for the time being, despite their perception as a marginalized group. What this chapter has shown is how the reality on the ground of what might simply be described as a city in conflict is, in fact, constructed by complex political-spatial processes.

Notes

1 At the time this study was carried out, from September 2011 to December 2015, Karachi was divided into five large administrative districts: Central, East, Malir, South and West. Published census and electoral data has been aggregated at this scale.
2 In the 2002 LG elections, MQM took 100 per cent of the seats in District Central and 75 per cent in District East, while in the 2008 LG elections they took 100 per cent of the seats in both districts.

References

Ahmed, F. (1998) *Ethnicity and Politics in Pakistan*. Karachi: Oxford University Press.
Allegra, M., Casaglia, A., and Rokem, J. (2012) The political geographies of urban polarization: a critical review of research on divided cities. *Geography Compass*, 6(9): 560–74.
Ansari, S. (1995) Partition, migration and refugees: responses to the arrival of Muhajirs in Sind during 1947–48. *South Asia: Journal of South Asian Studies*, 18(s1): 95–108.
Ansari, S. (2005) *Life After Partition: Migration, Community and Strife in Sindh 1947– 1962*. Karachi: Oxford University Press.
Anwar, F. (2010) Land use planning for unsustainable growth: assessing the policy to implementation cycle. *Commercialisation of Roads in Karachi: A Case Study*. Karachi: Shehri-Citizens for a Better Environment.
Charalambous, N., and Hadjichristos, C. (2011) Overcoming division in Nicosia's public space. *Built Environment*, 37(2): 170–82.
Gayer, L. (2014) *Karachi: Ordered Disorder and the Struggle for the City*. London: Hurst & Co.
Gazdar, H., and Mallah, H. B. (2013) Informality and political violence in Karachi. *Urban Studies*, 50(15): 3099–115.
Hasan, A. (1999) *Understanding Karachi: Planning and Reform for the Future*. Karachi: City Press.
Hasan, A. (2005) The political and institutional blockages to good governance: the case of the Lyari Expressway in Karachi. *Environment and Urbanization*, 17(2): 127–41.
Inskeep, S. (2011) *Instant City: Life and Death in Karachi*. New York: Penguin.
Jones, J. (2007) The Shi'a Muslims of the United Provinces of India, *c*.1890–1940. Pembroke College, University of Cambridge PhD.
Kershen, A. (2005) Mother tongue as a bridge to assimilation? In A. Kershen (ed.), *Stranger, Aliens and Asians*. Aldershot: Routledge: 133–65.
Khan, T. (2013) Cooking in Karachi: the world's most dangerous megacity is the next frontier in the global meth trade. *Foreign Policy*, 3 September. Available from: http://foreignpolicy.com/2013/09/03/cooking-in-karachi/. Accessed September 2015.
Legeby, A. (2009) Accessibilty and urban life: aspects on social segregation, in D. Koch, L. Marcus and J. Steen (eds), *Proceedings, Seventh International Space Syntax Symposium*. Stockholm: Royal Institute of Technology, 064: 1–11.

Maher, M. (2013) Fall sick with the healing: the 'killer' history of Karachi's first flyover *Express Tribune*, 2 December. Available from: http://tribune.com.pk/story/639770/fall-sick-with-the-healing-the-killer-history-of-karachis-first-flyover/. Accessed February 2014.

Mahmood, S. (1999) 'Shelter within my reach': medium rise apartment housing for the middle income group in Karachi, Pakistan. Master's Dissertation, MIT, Cambridge, MA.

Peach, C. (1996) Good segregation, bad segregation. *Planning Perspectives*, 11(4): 379–98.

Phillips, D. (2006) Parallel lives? Challenging discourses of British Muslim self-segregation. *Environment and Planning D: Society and Space*, 24(1): 25–40.

Rahman, T. (2011) *From Hindi to Urdu: A Social and Political History*. Karachi: Oxford University Press.

Robinson, F. (2014) Introduction: the Shi'a in South Asia. *Journal of the Royal Asiatic Society*, 24(3): 353–61.

Sennett, R. (1996) *Flesh and Stone: The Body and the City in Western Civilization*. London: Faber and Faber.

Shirlow, P., and Murtagh, B. (2006) *Belfast: Segregation, Violence and the City*. London: Pluto.

Sibley, D. (1995) *Geographies of Exclusion*. London: Routledge.

Siddiqi, A. R. (2008) *Partition and the Making of the Mohajir Mindset: A Narrative*. Karachi: Oxford University Press.

Simone, A. (2013) Cities of uncertainty: Jakarta, the urban majority, and inventive political technologies. *Theory, Culture and Society*, 30(7/8): 243–63.

Vaughan, L., and Arbaci, S. (2011) The challenges of understanding urban segregation. *Built Environment*, 37(2): 128–38.

Vaughan, L., Chatford Clark, D. L., Sahbaz, O., and Haklay, M. (2005) Space and exclusion: does urban morphology play a part in social deprivation? *Area*, 37(4): 402–12.

Verkaaik, O. (2004) *Migrants and Militants: Fun and Urban Violence in Pakistan*. Princeton, NJ: Princeton University Press.

Vertovec, S. (1995) Hindus in Trinidad and Britain: ethnic religion, reification, and the politics of public space. In P. van der Veer (ed.), *Nation and Migration: The Politics of Space in the South Asian Diaspora*. Philadelphia: University of Pennsylvania Press: 132–56.

Waterman, S., and Kosmin, B. A. (1988) Residential patterns and processes: a study of Jews in three London boroughs. *Transactions of the Institute of British Geographers* (NS) 13: 75–91.

Werbner, P. (2005) The translocation of culture: 'community cohesion' and the force of multiculturalism in history. *The Sociological Review*, 53(4): 745–68.

Zaman, F. (2013) Eight most violent flashpoints in Karachi. *DAWN.com*, 4 September. Available from: www.dawn.com/news/1040495. Accessed June 2015.

Zaman, F., and Ali, N. S. (2013) Taliban in Karachi: the real story. *DAWN.com*, 31 March. Available from: www.dawn.com/news/799118/taliban-in-karachi-the-real-story. Accessed June 2015.

5 The practice of 'marketplace coordination' in Jakarta (1977–98)

Pawda F. Tjoa

Introduction

This chapter critically analyses the politics behind local government's continuous attempts to organize the markets in the city of Jakarta under the banner of 'marketplace coordination', and uncovers the social tensions generated as a result of the top-down imposition of the policies associated with this strategy. Marketplace coordination denotes the multiple efforts by the government to impose a level of 'order' across the city's formal and informal market activities. This chapter will shed light on the construction of internal categories within the urban geopolitical process shaping the marketplaces of Jakarta.

'Marketplace' here refers to the physical sites in which goods are exchanged through the traditional means of haggling. These include both places located within a permanent structure and those that are situated in more temporary open-air spaces. The latter type of marketplaces contrasts markedly with more formal markets, in which the culture of haggling – characteristic of the traditional markets – is strictly prohibited by marketplace managers. While formal marketplaces dominate the neoliberal, commercialized 'shopping mall' culture of contemporary Jakarta, they have not replaced the traditional ones integral to the everyday life of the majority of Jakarta's population. Indeed, these two types of marketplace have coexisted since the opening of the first department store, Sarinah, in 1965. Indonesia's first president, Soekarno, intended to remove the uncertainties inherent in traditional markets through the fixed-price modern shopping centre. Nevertheless, attempts by the government to create an image of progress for Indonesia's capital city have implied the promotion of commercial structures amid the active marginalization of the traditional marketplace. Government support for formal commercial shopping centres kick-started a new trend of the regional mall, reminiscent of a similar process that occurred in 1950s Los Angeles.[1]

Despite the explosive moments that occurred during the implementation of urban policies under the marketplace coordination strategy, this period was known to be a relatively politically stable and peaceful time, which many scholars have attributed to the heavy policing and surveillance that characterized the regime under Indonesia's second president, Suharto, known as the New Order (1966–98) (Kusno, 2010: 2013; Colombijn, 2010: 116).

Ideology of development, stability and order

Between the 1960s and 1980s Jakarta's population grew to about 6.1 million and, by 2000, the population of the Greater Metropolitan Region reached 23 million (Hamnett, 2011: 248–9). Jakarta was becoming increasingly difficult to govern, despite the general impression of 'unity in diversity' portrayed by the state (Goldblum and Wong, 2000: 31). The first comprehensive master plan of Jakarta was the *Rencana Induk Kota* (*RIK*) (City Master Plan) 1965–85, which set out in detail clear intentions for future urban development in the capital city. Following this, a 20-year plan known as *Rencana Umum Tata Ruang* (*RUTR*) (General Spatial Plan) 1985–2005 was introduced "for the sake of [continuing] the existing interventions proposed in the *RIK*, and from which many of the internal conflicts described in this chapter emerged" (*RUTR*, 1985–2005: 1).

Following the governorship of Ali Sadikin (1966–77), Jakarta was managed by a series of comparatively short-lived governors under the New Order regime, which lasted a total of 32 years from 1966–98, and left visible imprints on the urban fabric of the capital city.[2] This chapter will focus on the intervention of Governor Tjokropranolo (1977–82), who succeeded Governor Ali Sadikin in 1977.

Drawing from Adorno's *Negative Dialectics* (1973), this chapter defines 'ideology' beyond its positivist terms – that is, to unite society – and, instead, reveals its potential to fragment society through the same set of doctrines. Ideology is broadly defined as a set of doctrines that determine what a 'good' society is and how to go about achieving it. However, Adorno (ibid.) suggests that the same 'ideology' that helps create a 'good' community can also create disunity and tension by virtue of its inherent characteristic of differentiation and 'othering'. In this chapter, I focus on policies related to marketplace coordination to reveal how the ideology of development, stability and order – despite being used by the state to create unity – exacerbated social fragmentation in this setting.

Creating stability through urban space transformations

One of the lasting legacies of the post-independence regime in Indonesia (1945–66) was a series of national building projects by President Soekarno. These projects were known as the *Mercu Suar* (lighthouse) because they were intended to constitute a 'beacon' of development and pride for the Indonesian people in the urban spaces of Jakarta. As shown in Figure 5.1, these projects comprised a series of architectural interventions, scattered across the Protocol Avenue as a reference to the Declaration of Independence in August 1945. This chapter will focus on

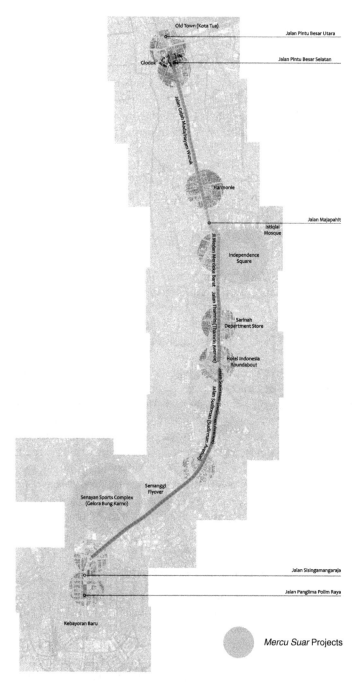

Old Town (Kota Tua)

Jalan Pintu Besar Utara

Jalan Pintu Besar Selatan

Glodok

Jalan Gajah Mada/Hayam Wuruk

Harmonie

Jalan Majapahit

Istiqlal
Mosque

Jl. Medan Merdeka Barat

Independence
Square

Jalan Thamrin (Thamrin Avenue)

Sarinah
Department Store

Hotel Indonesia
Roundabout

Jalan Sudirman (Sudirman Avenue)

Semanggi
Flyover

Senayan Sports Complex
(Gelora Bung Karno)

Jalan Sisingamangaraja

Jalan Panglima Polim Raya

Kebayoran Baru

Mercu Suar Projects

Figure 5.1 Mercu Suar projects Protocol Avenue and 'economy roads'
Source: Author, 2016

marketplace interventions under this rubric, which took place in conjunction with the development of the Sarinah department store, and the implementation of the city's first comprehensive master plan.

Sarinah

The ideological agenda of the 'Old Order' government under Soekarno was manifested in the construction of Sarinah in the 1960s, in the heart of Jakarta, as shown in Figure 5.1. Sarinah was used to symbolize 'stability' and 'order'; it offered the standardization of prices, thus removing the uncertainty found in traditional marketplaces. Meanwhile, the informal autonomous markets gradually came to be seen in negative terms – characterized as *kampungan* ('backward/ rural in character') – they were also described as exploitative of consumers, who were seen to be given neither control over nor information on the prices of goods. Sarinah, therefore, was introduced to act as a price stabilizer, providing an appropriate 'modern' solution to problems of informality, which would provide certainty and ensure 'fairness'. Nonetheless, to meet their daily household needs a majority of the population still relied on autonomous informal markets in which major interventions were also taking place in accordance with the new city master plan.

Master plans: RIK and RUTR

In the process of organizing marketplaces in Jakarta, a main avenue was privileged at the expense of less visible and prominent parts of the city. The main avenue is an urban armature stretching from the northern Old Town to the satellite city of *Kebayoran Baru* in south Jakarta. Drawing from Shane (2011: 37), this urban armature acts as a hierarchical "linear spatial organising device", constituting an important vehicle for driving southward while simultaneously representing the progressive modern image of the nation. If the 1960s and 1970s were remembered as a time of transition from the 'Old Order' to the 'New Order', during which the city underwent tough government 'ordering' policies such as the implementation of marketplace coordination, the reinforcement of traffic laws and the eradication of unruly trishaws from the streets of Jakarta, the 1980s and 1990s were generally remembered as a stable period during which these 'ordering' measures became part of the everyday, ingrained in the attitudes of the urban population.[3] The 'ordering' policies introduced during this time including the *Penertiban Ibukota* (Ordering of the Capital City), marketplace coordination and a series of *operasi* (operations) across Jakarta.

Marketplace coordination and the PD Pasar Jaya

The *RIK* (1965–85) recognized that there was a rising demand for conveniently located markets and commercial centres and the continued growth of the informal markets suggested that a persistent demand for their services was not being fulfilled by the formal amenities in the city (ibid.: 7). Therefore, the coordinating body *Perusahaan Daerah Pasar Jaya (PD Pasar Jaya)* was established in 1966 to help oversee and monitor the development of markets in Jakarta. As a regional enterprise, *PD Pasar Jaya* prioritized the generation of regional revenue to be invested back into the region.[4]

The establishment of *PD Pasar Jaya* purportedly helped to mediate the conflicts and tensions that occurred as a result of the lack of official contracts and documentation among independent merchants. However, the documentation carried out by *PD Pasar Jaya* – creating records of previously invisible merchants – and the simultaneous formalization of stall usage and, later, stall ownership highlighted the gap between social groups within the marketplaces. Thus, even though the *PD Pasar Jaya* was a regional enterprise, it largely acted as an extension of state bureaucracy, making these otherwise invisible activities more apparent to those in power.

Types of merchants and marketplaces

The market scene in Jakarta is often characterized by 'wild' petty merchants who are familiarly known as *pedagang kaki lima* (*PKLs*). By September 1979, the number of *PKLs* in Jakarta Capital Special Region (DKI Jakarta) had reached about 100,000 (Marzuki Arifin 348, 6 September 1979). The markets, managed by the *PD Pasar Jaya*, were not limited to the *PKLs*, but included others located within a permanent market building. However, due to the temporary nature of the *PKLs* they were more prone to *pungutan liar/pungli* (exploitative 'wild' remuneration collectors).

Meanwhile, the merchant communities were diverse, consisting of people who lived in Jakarta and others who travelled to the capital city during special seasons or festivities. Aside from this distinction, merchants were also often categorized into *pribumi* (native) and non-*pribumi*; these categories were frequently used to distinguish financially weaker merchants from well-off ones.

Funding structures

While most markets emerged independently and became gradually absorbed by the *PD Pasar Jaya*, others relied on the funding provisions within the *PD Pasar Jaya*. Some markets, overseen by the *PD Pasar Jaya*, are known as *Pasar-Pasar Inpres* (*Inpres* markets), as they are funded by a specific fund known as *Dana-Dana Inpres* (*Inpres* Fund) that sought to promote 'national entrepreneurship'.[5] The main purpose of this funding scheme was to promote entrepreneurial attitudes among the population of Jakarta, particularly those who were considered *berekonomi lemah* (financially weak). The stalls in *Pasar Inpres* were only to be rented out on a daily basis, and not to be sold.

Additionally, poorer merchants relied on a lottery system to gain access to market stalls. This lottery system was originally established in order to give *PKLs* the opportunity to set up their business in a number of prime locations across

Jakarta, including in places like *Pasar Inpres Senen* and *Pasar Inpres Johar Baru* in Central Jakarta.[6] However, this lottery system is often manipulated by the authorities, resulting in a widespread culture of bribery. The *PD Pasar Jaya* quickly gained a reputation for corruption, and was even likened to the *pungutan liar/pungli* ('wild' collectors/mafia/*preman*) who still controlled a majority of the independently managed markets outside of the *PD Pasar Jaya*.[7]

In May 1979 the *PD Pasar Jaya* officially managed only 19 *Pasar Inpres*, while a majority of the markets in Jakarta were still managed independently. By this time, the collection of remuneration by the authorities had spiralled out of control, and a number of markets managed by the *PD Pasar Jaya* were left significantly empty and unused. The authorities' fee collection became increasingly suspicious, without any clear administrative protocol, resulting in inconsistent revenue reports. There was also evidence suggesting that the leaders of the markets had demanded unlawful remuneration/bribes.[8] Although the widespread practice of bribery and corruption was recognized quite early on as one of the major hurdles in the successful operation of the *PD Pasar Jaya*, the level of intervention in tackling it was insufficient.[9]

Purpose and governance

In his thesis on the Thamrin-Sudirman Avenue, Kusumawijaya (1990: 7-1) argues that the principles guiding the development of this main avenue has shifted from political to economic ones, even if the morphological value remains significant. He further concludes that the avenue's markedly larger scale enabled the urban armature to be 'reused' repeatedly by another regime to fulfil a new agenda for the city. The highly ideological and idealist development of the urban armature in the 1950s and 1960s was arguably the first post-independence regime's biggest contribution to the symbolic power of Jakarta as Indonesia's the new capital city; this morphological transformation, and the aesthetics chosen by the first president, Soekarno, through and for the *Mercu Suar* projects, remained versatile, constantly being manipulated to suit a series of radically new ideologies. The *Mercu Suar* projects were aimed at building national identity and pride for the Indonesian people. These would include a series of monumental buildings designed according to the modernist paradigm; the International Style, in particular, was chosen as a way of disassociating the independence regime from any remnants of the colonial period. In this way, the main aim of these monumental projects was not so much to fulfil the many practical needs of the population, but to construct an image of progress that the ruling elite deemed essential for establishing a national identity.

The urban armature remains dominant both symbolically and at an urban scale, and in this way it both fragments and links together different parts of the city, as well as multiple social groups within Jakarta. As also previously suggested by Kusumawijaya (1990: 7-3), the avenue "collects the elements along the area that it passes". In the remaining parts of the chapter, we will see how this pursuit of order and the image of stability materialized specifically through the marketplace coordination effort, which perpetuated alienation and urban fragmentation among existing communities.

Challenges of marketplace coordination: indoor retreat and displacement of PKLs as a signal of progress

The activities of the *PD Pasar Jaya* frequently caused great anxiety among the merchant communities, who used to operate with great autonomy, meaning that they were accountable to neither the authorities nor any coordinating bodies. In fact, most merchants were largely absent from the state's official records. The common strategy, employed by the authorities, of relocating *PKLs* into indoor market spaces was, arguably, a significant trigger for a generalized retreat indoors. In a study of Jakarta's marketplaces during the colonial period, Kusno (2010: 189–92) argues that the construction of dedicated bazaars to contain other-wise 'unruly' informal markets were intended for the "construction of normality in the urban life". Activities conducted outdoors increasingly became perceived as rebellious and indecorous to the state's dominant ideology of order, stability and development. The *PD Pasar Jaya* thus further encouraged a form of urban retreat from the main streets and avenues of the capital city. Even though these same avenues, as in Figure 4.1, were specifically labelled as high-intensity commercial, industrial and office zones (indeed, many were known as 'economy roads'), this label excluded the *PKLs*. This exclusion disregarded their critical role in the daily operation of the city. The *PKLs* were removed from urban space and eliminated from the urban spectacle because they were seen to taint the ideal image of a modern commercial hub. This reflected Drummond's (2000: 2380) observation, in the context of Vietnam, that urban public space is frequently understood as a site that allows marginalized actors to challenge the dominant order. In Jakarta, however, the marginalized were excluded from the ideal urban spectacle of the city, in which urban space was perceived as "a controlled and orderly retreat where a properly behaved public might experience the spectacle of the city" (ibid.: 2379–80). It is in this same way that marketplace coordination exacerbated the practices of alienation evidenced in rising socio-economic and ethnic tensions in the merchant community in Jakarta.

Kusumawijaya (1990: 7-4) argues that, with the privatization of the plots on either side of Thamrin-Sudirman Avenue, the roads came to be regarded as a negative or leftover space, rather than an architectural element that could be designed and carefully integrated into Jakarta's existing urban fabric. This meant that the urban armature was treated not as a linear connecting device, which 'collected' and brought together different parts of the city, but merely as a means of travelling from point A to point B. 'Collects' is a term used by Kusumawijaya (1990) in reference to the metropolis. For Kusumawijaya (2006: 120) this north–south axis comprises Jakarta's most important feature: it acts as a collective memory. It is morphologically important because it makes Jakarta legible; and it is functionally central, because it makes the metropolis come to life. In effect, he sees the north–south axis as a main artery, pumping blood into the city. However,

Jakarta's notorious problems with traffic congestion mean that, despite this artery, one does not usually enjoy a straightforward commute. A recent study suggests that Jakarta's population spends on average two hours a day in traffic, and, in 2015, it was ranked the most congested city in the world (*The Jakarta Post*, 9 February 2015, 5 February 2015). With the campaign to keep people off the streets through the indecorous association attached to activities conducted in the urban outdoors, it seems as though those who insisted on remaining outdoors have been punished by being permanently stuck in stationary traffic.

This is not the only irony relating to the urban experience of Jakarta. The observations made about the city's marketplaces offer inconsistent information about the adequacy of the market spaces for merchants' needs. At first glance, the problem of vacant market buildings indicates an oversupply of market space; yet *PKL*s were often still found visibly overflowing into the streets.

Pungutan liar/pungli ('wild' collections)

The imbalance in the supply and demand of stalls and kiosks in the markets arguably promoted the practice of *pungutan liar/pungli* (wild collections by the 'mafia'/*preman*). Merchants who did not manage to secure a stall/kiosk began to sell their goods along the edges and by the entrances to formal marketplaces. They often overflowed into the roads and various access points in the city, where they had to negotiate and compete with the usage of the same spaces by moving vehicles. Nevertheless, constantly under the threat of removal, the *PKL*s usually paid the *pungli*/mafias/*preman* 'protection money' to guard them against eviction by the authorities.[10] In this way, the *PKL*s constantly inhabited the peripheral 'in-between' spaces, which Simone and Rao (2012: 331) refer to as "spaces of uncertainty" which, in turn, may generate spaces of innovation.[11] Given the lack of coordination by the authorities, the *pungli*'s assurance to the *PKL*s was as good as an unofficial right to remain.

The inability to control the *PKL*s made it necessary for the authorities to reach a series of compromises, including ignoring their existence, which, in turn, granted the *PKL*s an ambivalent status that enabled them to thrive on their own. However, this ambivalence exposed the *PKL*s to continued manipulation by the *pungli*. In *Pasar Tanah Abang*, the problem of *pungli* was particularly rampant and well documented, with the *pungli* receiving an income of IDR 10 million (US$16,000).[12]

Concerned about the widespread 'wild' collection of 'security' remuneration (also known as protection money), the governor mobilized a comprehensive crackdown. Further investigation remarkably revealed that the 'wild' collection was conducted with the backing of an *Angkatan Bersenjata Republik Indonesia* (ABRI, Indonesian National Armed Forces) official (Marzuki Arifin 273, 18 August 1979).

Despite existing prohibitions against trading in public spaces, the Governor decided to be lenient to the *PKL*s who broke this law, especially during the period approaching the *Lebaran* (Eid festivities) (Marzuki Arifin 348, 21 August 1979).

This was thought to prevent the continued exploitation of the *PKLs* by the *pungli*. The practice of *pungli* became so rampant in the late 1970s that it had to be tackled by both the *Himpunan Pengusaha Pribumi Indonesia* (HIPPI, Indonesia's Indigenous Entrepreneurs Union) and the *Kodak Metro Jaya* (regional police force).

The complicit nature of not seeing was, however, a hindrance to the modernist agenda of the state. As Cowherd (2008: 284) once noted, the "depth and inertia of the New Order ideology is such that even those without cars see the removal of small-scale merchants, bicycle taxis and pedestrian space as signals of progress". These signals were welcomed because they pointed to greater stability and order. Indeed, it was this ideology of development that continued to displace street merchants, resulting in substantial retreat from urban areas in Jakarta. Nevertheless, this too was popularly regarded as a sign of progress. The placement of the *PKLs* into a category of deviance to be contained within dedicated marketplaces across Jakarta turned out to be extremely problematic, since it resulted in tension among merchants themselves, as well as between the authorities and the affected merchants.

The conflict generated between different groups of people as a result of the marketplace coordination effort by the *PD Pasar Jaya*, however, became the ultimate ingredient for the creation of a more relevant form of public space in Jakarta. The markets acted both as a 'space of appearance' and a 'common world', which according to Hannah Arendt are the two necessary conditions for the creation of a public space. For Arendt (1958: 46), these two dimensions are distinct from one another but interrelated. The first, the space of appearance, provides political freedom and equality made alive through citizens acting together through speech and persuasion. The second, the common world, is a shared man-made institution or artefact that sets humanity apart from other beings (or nature) and provides us with a permanent venue for our activities. In Indonesia, parallels can certainly be discerned between the previous colonial political order and more recent urban development. Cowherd (2008: 275–86) has highlighted the 'heterotopian logic of the colonial order', which provided a firm foundation for the reproduction of a similar disparity in power relations and social dualism.[13] In the sense of a 'post-civil society', the markets created and managed by the *PD Pasar Jaya* could be considered a heterotopia, which may be understood to re-emerge as "a strategy to reclaim places of otherness on the inside of an economised public life" (Dehaene and De Cauter, 2008: 4). Indeed, the marketplaces created under the guidance of the authorities can be perceived as a specialized urban element (or enclave), with multiple interior subdivisions, which is able to hold conflicting urban activities at the same place and time. Much like Foucault's (1984: 46–9) "heterotopias of deviation", which include prisons, hospitals, clinics, asylums and courthouses, these marketplaces were created precisely to house people whose activities did not fit within the ideal image of the city. In Jakarta, merchants and authorities alike attempted to make sense of the ways in which the marketplace operated through a number of strategies of reclaiming 'in-between' spaces of uncertainty. Indeed, we observe within the marketplaces the tension that Dehaene and De Cauter (2008: 5) felt to be embodied within the heterotopia – "between place and non-place – that continued to refashion the nature of public space". As detailed in the following section, these strategies were also pursued in the development of further internal social categories, and the promotion of market stall ownership.

Market stall ownership

The discrepancies between the everyday requirements of the merchants conducting trade and the efforts on the market management to fulfil these needs too often reflected a common symptom of many failed planning efforts in Jakarta (Marzuki Arifin 348, 9 July 1982). In *Pasar Senen*, the market management went to the extent of waiving rental fees in order to encourage the use of vacant stalls. However, this effort was in vain, as the *PKL*s refused to relocate to an empty market building even after rental fees had been waived (Marzuki Arifin 348, 10 July 1982). While stall ownership – that offered security and stability – was appealing for the 'permanent' merchants in the area, it was not attractive for *PKL*s, who preferred to remain mobile. Nevertheless, stall ownership inadvertently distinguished the wealthier merchants from their poorer counterparts.

The *pasar-pasar* (marketplaces) created under the guidance of the authorities can be perceived as a 'heterotopia of deviance' – a specialized urban element (or enclave), with multiple interior subdivisions, that may contain conflicting urban activities at the same place and time. Shane (2011: 14–16) borrows the term from Foucault, who determined prisons and hospitals, among others, as "heterotopias of deviation", created precisely to house people who do not fit within the ideal image of modernity. The same strategies used to remove beggars from the streets of Jakarta were used to deal with the *PKL*s. Figure 5.2 shows the unruly *PKL*s commonly seen in the city, occupying what should be a public vehicular circulation. Even though beggars and petty traders were distinct in character, when situated in the context of the pursuit of an ideal image of the city, they were both placed under a larger category of deviance. By compartmentalizing and removing people who fell within this category, the constructions of 'heterotopias of deviance' would facilitate the production of the modern city.[14]

Throughout the 1980s, the remuneration-paying merchants and stall-owners complained that they were at a significant disadvantage, due to the presence of unfair competition by the *PKL*s. The *PKL*s not only caused considerable disorder within the general appearance of the markets but also deprived the legitimate merchants of potential customers. However, despite repeated attempts to relocate the *PKL*s into permanent market stalls, the number of *PKL*s, particularly during *Lebaran*, was so large that during this specific season the market management had to open up the parking lots in *Jatinegara* and *Pasar Mayestik* in order to accommodate them.[15]

The market coordination effort to relocate the *PKL*s and other merchants into indoor spaces to achieve an appearance of order effectively displaced street

Figure 5.2 Unruly petty traders blocking traffic
Source: Marzuki Arifin Archives 348 'Menjelang Lebaran Razia Dikendorkan', August 1979. Courtesy of the National Archives of the Republic of Indonesia (ANRI)

vendors and provided an opportunity for a collective identity to be created – albeit one that differentiated one group from another (Kusno, 2000: 105). As merchants began to be categorized according to capital value, wealth, ethnicity and other socio-economic indicators, the distinctions between groups became increasingly obvious. For instance, those who agreed to be relocated into official kiosks within the market had to abide by certain rules, and pay rents and remuneration accordingly. Importantly, their participation in this urban relocation effectively placed them in a superior category of 'remuneration-paying' merchants, as opposed to the 'wild' *PKLs*, who generally came and went as they pleased. The latter could then be blamed for having an unfair competitive advantage over the 'remuneration-paying' merchants, who did not possess the same level of flexibility. The remuneration-paying merchants, accordingly, began to set themselves apart from their apparently 'disruptive' counterparts. Despite their flexibility, *PKLs* were faced with the constant threat of eviction at any time without warning.

Furthermore, the ownership of stalls became popular in the 1980s and was, indeed, promoted by the *PD Pasar Jaya*. The ability to own stalls and the prospect of stall ownership presented a huge contrast to the rent-only restrictions imposed on merchants just 30 years prior.[16] However, the right to own became a point of contestation, as it turned what used to be a traditional shared market into an increasingly privatized space, which was difficult to monitor and regulate. One significant consequence of this lack of regulation and monitoring was the deviation of some stalls from their intended functions, arguably for the sake of economic survival – the alternative was difficult to accept, since market stalls that failed to adapt quickly to changing tastes often became disused or obsolete. Examples of this appropriation could be found in *Pasar Tanah Abang* and *Pasar Tugu*, in which stall-owners decided to stop using stalls as kiosks, due to a lack of regular customers frequenting the markets. Instead, they turned the stalls into mini warehouses acting as an intermediate distribution centre for more productive market kiosks elsewhere. Many stall-owners had even been able to store their merchandise within the tiny two-square-metre stall space while also conducting wholesale transactions from within the same space. In spite of these minor violations, whereby stall-owners altered the intended use of the marketplace, *Pasar Tanah Abang* gained a reputation as the resilient textile hub/headquarters of Jakarta that it retains to this day (Marzuki Arifin 348, 13 July 1982).

This emphasis on stall ownership also further highlighted ethnic divisions within merchant communities. In January 1981, existing merchants of *Pasar Inpres Duren Sawit* protested when the market management decided to allow non-*pribumi* merchants to set up business in the markets, which until then had been exclusively available to the *pribumi*. The *pribumi* merchants were concerned that the area would become gentrified as more stalls were being purchased by non-*pribumi* merchants, quickly overtaking the number of original *pribumi* merchants with neither the financial resources to purchase stalls nor the means to survive against fiercer competition. Merchants in markets such as *Pasar Inpres Duren Sawit* felt threatened not only by the cultural change in the existing merchant communities, but also by the profit-oriented aims of the *PD Pasar Jaya* – a significant deviation from the original aims of the *Pasar Inpres*.

The profit-oriented approach of the management *PD Pasar Jaya* gradually stripped the traditional markets of the most essential characteristics of a public space, thus alienating the market culture from the core character of the everyday life of the ordinary inhabitants of Jakarta. While many stalls that had been bought up remained vacant, the *PKLs* and other forms of informal activities continued to attract customers to the peripheries of the partially vacant market buildings. The challenge that the government and planners continued to face was recognizing the

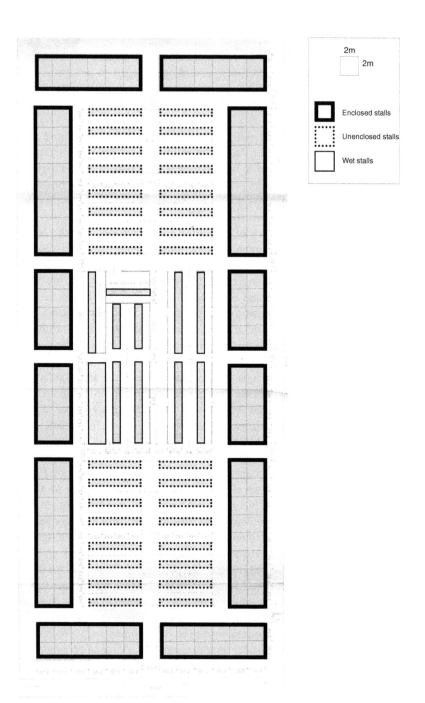

2m

2m

Enclosed stalls

Unenclosed stalls

Wet stalls

Figure 5.3 Plan of *Pasar Mayestik Blok 1*

Source: Marzuki Arifin Archives 160, 168 and 323. Courtesy of the National Archives of the Republic of Indonesia (ANRI)

valuable contribution of the *PKL*s, while at the same time addressing the practical problems arising from their continued operations in many parts of the city.

Market layout

Another example of the authorities' failure to recognize the fundamental charac-teristics of the *PKL*s could be seen in the attempt to relocate them indoors through the *Inpres* Fund. This strategy failed to take mobility as a central feature of *PKL*s; to most, the prospect of having a permanent stall was not, in fact, appealing. Indeed, the typically opaque market layout with 'closed' stalls along the outside walls, as shown in Figure 5.3 of *Pasar Mayestik Blok 1*, made attracting custom-ers extremely ineffective. These markets, therefore, did not at all reflect the 'porous' nature of the traditional markets seen earlier in Figure 5.1. The rigid spatial layout of the opaque market layout that does not allow a more fluid and free flow of interactions appealed to neither merchants nor customers.

Additionally, the obligation to give remuneration was seen as a significant disincentive by many *PKL*s, who preferred to continue with the uncertainties associated with being informal mobile vendors. By the end of 1978, the highest concentrations of *PKL*s across the five districts in DKI Jakarta were found in 712 key locations, including *Tanah Abang, Pasar Kebayoran Lama, Blok M-Melawai, Pasar Bendungan Hilir* and Sudirman Avenue, 20 of which had a density of over 500 *PKL*s on site.[17] Meanwhile, the practically non-existent customer base in the formal marketplaces was deeply demoralizing as 25 per cent of the total stalls/ kiosks had to be consequently closed down. The persistent presence of the *PKL*s along the edge of the markets continued to draw valuable customers away from the remuneration-paying and law-abiding merchants within.

Conclusion

Marketplace coordination consisted of the creation of heterotopic structures that – especially between the 1970s and the 1990s – contributed to the mentality of alienation among Jakarta's population. Through the *PD Pasar Jaya*, we see glimpses of the government's attempt to compromise their urban geopolitical policies to work with the persistent *PKL*s across Jakarta. It is vital to note the perennial struggle to persuade these 'wild' merchants to utilize the stall spaces provided in the *Pasar Inpres*. It was also clear that the main reasons for the empty stalls and markets were their poor location and design, which failed to reflect the 'porous' nature of traditional markets. Importantly, during the months and weeks approaching the *Lebaran*, when Jakarta was inundated with seasonal *PKL*s in strategic but often inappropriate locations (e.g. by traffic lights and the busiest roads in the city) the local government was forced to become lenient. In many instances, the authority had to accommodate the needs of the seasonal *PKL*s who had come from the villages into the capital city during the *Lebaran* high season. The government realized that it would have been impossible to impose the stand-ard rules on the *PKL*s during this busy period; a full-scale eviction might result in an even greater chaos or even violence, a risk too much for the government to endure.

It may seem tenuous to suggest that the markets are a form of 'heterotopia of deviance', but, on closer examination, we see that they performed the same function as the rehabilitation centres and prisons for the prostitutes and the criminals. First, the markets were intended to become a collection point for the chaotic and uncoordinated *PKLs*, in the same way that the rehabilitation centres were intended to house prostitutes, beggars and the homeless, keeping them out of sight of the streets of Jakarta. These categories each, in their own way, gave the government a heightened sense of anxiety and uncertainty that was characteristic of a country in pursuit of everything 'modern'. Of course, marketplaces were very much integrated as part of the economic processes of the city, whereas the rehabilitation centres may not necessarily have as distinct an economic link to events happening in Jakarta. However, the ultimate aim of the rehabilitation centres in 1960s and 1970s Jakarta, which was to transform prostitutes into a productive labour force that conformed to the moral and ideological principles of the state, meant that they were seen as a potential part of the economic structure, capable of contributing to the economy of the city. In this way, the markets and the rehabilitation centres were not functionally that different after all.

Even though the persistence of the *PKLs*, especially during the *Lebaran* period, posed an enormous challenge, they turned the streets into a positive space, filled with mingling and stopping, instead of the traffic circulation for which it was originally intended. In fact, the streets became a rather familiar form of public space for most long-term residents of Jakarta. Just as special international events often stepped up efforts to create an ideal urban spectacle for the foreign visitors, a special occasion such as the *Lebaran* led to this same project, linked to the creation of a modern city, to be momentarily suspended.

Notes

1 See Longstretch, 1998: xiv.
2 'New Order' (*Orde Baru*) was a term coined by the second president of Indonesia, Suharto, to set his regime apart from the 'Old Order' under the first president, Soekarno. The New Order lasted from 1966–98, the longest-lasting post-independence regime.
3 The 'ordering' policies in the 1960s and 1970s included *Penertiban Ibukota* ('Ordering of the Capital City'), Ordering of the Markets through *PD Pasar Jaya* and a series of *operasi* ('operations').
4 This regulation on *PD* was based on the law passed in 1962. Both state-owned and privately owned companies could constitute a *PD* as long as it contributed to the regional revenue in taxes.
5 *Inpres* stands for *Instruksi President* ('Presidential Mandate'). The *Pasar Inpres* refers to marketplaces whose construction and maintenance were funded by reserves set aside for specific presidential mandates.
6 According to a document on the modernization of the *Mayestic Blok I* market by Sugiyono Cablaka, the lottery system was thought to cause huge dissatisfaction among the merchants; a *sistem penunjukan* ('nomination/appointment system') was seen as more appropriate.

7 See Marzuki Arifin Archives 348, 30 May 1979.
8 The leaders of the markets had been appointed to oversee the market activities.
9 See Marzuki Arifin Archives 348, 30 May 1979.
10 Note that while 'mafia' is usually associated with the Italian criminal phenomenon, this term is sometimes used in Jakarta – and, indeed, other cities in Indonesia – often interchangeably with *preman* ('gangsters'), to refer to a similar group that has great indirect economic and political influence on local communities, usually through their persistent underground operation.
11 See also Simone, 2010: 40, and Roy, 2011: 232.
12 This value is based on currency exchange rate in 1979 US$1 = IDR 626.994.
13 Also see Dehaene and De Cauter (2008: 8).
14 'Deviation' in this case refers to departure from a strict set of ideologies.
15 See Marzuki Arifin Archives 348, 2 July 1982.
16 The ownership of market stalls was initially strictly forbidden, to ensure that the markets were accessible to the wider group of merchants and not limited to those who could afford to buy the stalls.
17 See Marzuki Arifin Archives 348, 'Kaki Lima jadi pedagang formil, bukan sekedar pindah ke pasar'.

References

Abel, H. (n.d.) *Kaki Lima jadi pedagang formil, bukan sekedar pindah ke pasar* [newspaper]. ANRI (National Archives of the Republic of Indonesia), Marzuki Arifin Archives 348, Jakarta.

Adorno, T. W. (1973) *Negative Dialectics.* New York: Seabury Press.

Arendt, H. (1958) *The Human Condition*, 2nd edn. Chicago: University of Chicago Press.

Cablaka, S. (n.d.) *Rencana Peremajaan Pasar Pusat Mayestik Blok I* [document]. ANRI, Marzuki Arifin Archives 168, Jakarta.

Colombijn, F. (2010) *Under Construction: The Politics of Urban Space and Housing During the Decolonization of Indonesia, 1930–1960.* Leiden: KITLV Press.

Cowherd, R. (2008) The heterotopian divide in Jakarta: constructing discourse, constructing space. In M. Dehaene and L. De Cauter (eds), *Heterotopia and the City: Public Space in a Postcivil Society*. Abingdon: Routledge: 275–86.

Dehaene, M., and De Cauter, L. (eds) (2008) *Heterotopia and the City: Public Space in a Postcivil Society.* Abingdon: Routledge.

Drummond, L. B. W. (2000) Street scenes: practices of public and private space in urban Vietnam. *Urban Studies*, 37(12): 2377–91.

Foucault, M. (1984) Of other spaces, heterotopias. *Architecture, Mouvement, Continuité*, 5: 46–9.

Goldblum, C., and Wong, T.-C. (2000) Growth, crisis and spatial change: a study of haphazard urbanisation in Jakarta, Indonesia. *Land Use Policy*, 17(1): 29–37.

Gubernur ijinkan pengaplingan pasar Jatinegara untuk pedagang kakilima (1982) *Berita Buana*, 2 July [newspaper]. ANRI, Marzuki Arifin Archives 348, Jakarta.

Gubernur perintahkan selidiki pungli di pasar tanah abang (1979) *Kompas*, 21 August [newspaper]. ANRI, Marzuki Arifin Archives 348, Jakarta.

Hamnett, S., and Forbes, D. (2011) *Planning Asian Cities: Risks and Resilience.* Abingdon: Routledge.

Jakartans spend 400 hours a year in traffic, says survey (2015) *The Jakarta Post*, 9 February [online newspaper]. www.thejakartapost.com/news/2015/02/09/jakartans-spend-400-hours-a-year-traffic-says-survey.html. Accessed 19 September 2016.

Kios pasar tanah abang tetap dilarang digunakan untuk gudang (1982) *Berita Buana*, 13 July [newspaper]. ANRI, Marzuki Arifin Archives 348, Jakarta.

Kusno, A. (2000) *Behind the Postcolonial: Architecture, Urban Space, and Political Cultures in Indonesia*. London: Routledge.

Kusno, A. (2010) *The Appearances of Memory: Mnemonic Practices of Architecture and Urban Form in Indonesia*. Durham, NC, and London: Duke University Press.

Kusumawijaya, M. (1990) Thamrin-Sudirman Avenue: a case study in the problem of modernization in a developing metropolis. Master's thesis. Katholieke Universiteit Leuven.

Kusumawijaya, M. (2004) *Jakarta: metropolis tunggang langgang*. Yogyakarta: GagasMedia.

Kusumawijaya, M. (2006) *Kota Rumah Kita.* Jakarta: Borneo.

Longstretch, R. W. (1998) *City Center to Regional Mall: Architecture, the Automobile, and Retailing in Los Angeles, 1920–1950.* Cambridge, MA, and London: MIT Press.

PD Pasar Jaya gagal membina pasar2 di Jakarta (1979) *Sinar Pagi*, 30 May [newspaper]. ANRI, Marzuki Arifin Archives 348, Jakarta.

Pedagang K-5 musiman bikin pusing (1982) *Berita Buana*, 10 July [newspaper]. ANRI, Marzuki Arifin Archives 348, Jakarta.

Pedagang2 K-5 musiman bandel tidak mau ditertibkan baik2 (1982) *Tertib*, 9 July [newspaper]. ANRI, Marzuki Arifin Archives 348, Jakarta.

Pemerintah DKI Jakarta (1966) *Rencana Induk Jakarta 1965–1985.* Jakarta: Pemerintah DKI Jakarta.

Pemerintah DKI Jakarta (1984) *Rencana Umum Tataruang Daerah-Daerah Khusus Ibukota Jakarta.* Jakarta: Pemerintah DKI Jakarta.

Roy, A. (2011) Slumdog cities: rethinking subaltern urbanism. *International Journal of Urban and Regional Research*, 35(2): 223–38.

Rp 10 juta hasil pungli di pasar Tanah Abang (1979) 6 September [newspaper]. ANRI, Marzuki Arifin Archives 348, Jakarta.

Samodra, H. B., and Tanjung, Z. (1979) Pungli di Pasar Tanah Abang, 18 August. ANRI, Marzuki Arifin Archives 273, Jakarta.

Shane, D. G. (2011) *Urban Design Since 1945: A Global Perspective*. Chichester: Wiley.

Simone, A. M. (2010) *City Life from Jakarta to Dakar: Movements at the Crossroads.* London: Routledge.

Simone, A., and Rao, V. (2012) Securing the majority: living through uncertainty in Jakarta. *International Journal of Urban and Regional Research*, 36(2): 315–35.

Wardhani, D. A., and Budiari, I. (2015) Jakarta has 'worst traffic in the world'. *The Jakarta Post*, 5 February [online newspaper]. www.thejakartapost.com/news/2015/02/05/jakarta-has-worst-traffic-world.html. Accessed 19 September 2016.

6 The politics of *doing nothing*

A rethinking of the culture of poverty in Khulna *c*. 1882–1990

Apurba Kumar Podder

Introduction

The 1980s in Bangladesh was a period of crisis for the state. Throughout the decade, the state balanced the budgets, cut subsidies and sought international loans to adjust government budgets. The state had to adopt fiscal austerity, affecting its ability to provide support and services for the poor. This chapter will explore how, in this crisis, the evicted vendors of an illegal market in Khulna city, locally known as *Kacha-bazaar*, managed to occupy (unlawfully) a piece of public land and what empowered them to stand there throughout the decade, and see off the threat of eviction.

The exploration gives a unique glimpse into the spatial politics of the poor in one of the important coastal cities in Bangladesh, where the poor are considered overpowered and exploited by the local political leaders. The chapter stresses that particular traits of the poverty culture must be (re-)examined with respect to sustained exploitation to understand a new form or resilience of the poor in emergence in the region. The chapter discusses the act of *doing nothing* within a context of a city where political leaders maintain overwhelming control of the ways government decisions are implemented for the development of the city.

The politics of *doing nothing*

During the 1980s, a typical everyday image of the illegal *Kacha-bazaar* in Khulna, a mid-sized city in Bangladesh, was of a group of people sitting idly with cigarettes in their mouth, doing nothing. Men in theirs 30s are the common participants in this pastime: gossiping, playing carom or engaging in *adda* (informal discussion) in a shabby clubhouse-cum-office in illegal settlements. While doing nothing is part of the entertainment and breaks (from work), it is also seen as a waste of human labour and manpower. Its everydayness, however, makes it worthy of further examination to gain a critical understanding of the culture of poverty. As Arjun Appadurai (2004) points out, culture should be considered as a 'capacity' in which the ideas of the future and the past are embedded and nurtured, therefore a critical examination of such images is imperative for effective policy-making to change the lives of the poor.

Most often, doing nothing is seen as a forced condition, one in which the poor are made to wait for opportunities. Being socially, politically and geographically outside the hegemonic power structure, the poor are seen as restricted from accessing public resources and earning their basic means in the contested cities of South Asia. They were, according to this view, forced into unemployment by exclusionary developmental agendas, and made to do nothing until they could gain access to resources and opportunities. Craig Jeffrey (2010) observes that the poor or lower-middle classes are obliged to wait long periods of time by powerful institutions with particular visions of the future. Jean-François Bayart (2009) has argued that long-term experiences of 'waiting' have become a common feature of the experience of subaltern people across the world after the 1960s. Ethnographic research on asylum seekers, refugees, the unemployed and rural poor are full of references to support Bayart's argument, demonstrating how waiting or doing nothing is a forced condition in the everyday life of the poor in developing cities (Jeffrey, 2010).

An alternative scholarly tradition sees doing nothing as a cultural 'choice' to reclaim a certain form of dignity within a context of exploitation and discrimination. In the last two decades, a growing literature has ascribed overwhelming efficacy to electoral politics in South Asia to play on the lives of the urban poor. For example, Ananya Roy (2004) has explored the illegal squatter camps in Calcutta and shows how the poor are persistently exploited by both a (urban) developmental culture deeply structured by class logic and the electoral political culture of the state. Roy demonstrates how slum dwellers engage with that political culture, investing substantial amounts of their time in it. Some slum men describe themselves as being unpaid 'mobilizers' of people for a political party, 'maintaining the club' and unity. However, in Roy's (2004, 2011) work, such claims of contribution meant very little. Roy understands these people's supposed activities to be in effect an excuse to *do nothing* as she quotes the working women of the slums: "we work all day long and they stay at home. They become just like the *babus*" (Roy, 2004: 63). Their vague contribution to the informal settlement, in Roy's portrayal, is not only a veil of masculinity to hide their unemployed status, but also an ironic culture of poverty.

Roy suggests that the 'negotiability' inherent in the political culture of the state perpetuates the engagement of the poor, and, through such interaction with the state apparatus, the poor men feel a sense of 'self-importance' (ibid.: 161). Roy traces the roots of such behaviour to the class discrimination of rural land societies. The search for status witnessed among illegal settlers who occupy urban land has its roots in rural agrarian hierarchy. While their rural landlessness denied them access to rural political institutions, class-mediated urban politics was permeable and gave them access to party patronage. Drawing on Raewyn Connell (2005), Roy sees such doing nothing as 'marginalized masculinity', an ensemble of practices and discourses through which poor men locate themselves in class and gender hierarchies (Roy, 2004: 161).

Similar to Roy, Appadurai (2004) claims that observations of the poor in different parts of the world show the poor (to) maintain forms of irony and distance from the existing socio-cultural norms in ways that allow them to maintain some sense of dignity in conditions of oppression. The problem with such scholarly assessment is that it sees doing nothing through what Peter Townsend (1979) calls a 'lens' of middle-class values, and shares anxieties held by the state to distinguish 'deserving' and 'undeserving' poor. Especially under economic liberalization, which was imposed in the 1980s and firmly practised in the following decades, the assessment of who exactly constituted the 'deserving' poor depends on the ability of subjects to exhibit self-discipline. Such views essentially carry the legacy of colonial culture, defining 'work' by the principles considered necessary to sustain the market-driven economy. For example, in the face of labour shortage in England, the Vagrancy Law was passed in 1349 in order to discipline the poor who were doing nothing (Chambliss, 1964). The legislation was rooted in a variety of motivations, chief among them the need to address the severe labour shortage created by the plague and the migration of peasants to urban areas in search of improved living conditions (Bradshaw, 1921). The law was intended to force anyone who was able to work to do so. In such a situation, one's fundamental social value and contributions are judged by one's relationship with 'productive' work or the mode of production.

Aligning with these latter discourses in the literature, in this chapter I adopt a radical notion of 'doing nothing' by examining the ways in which idleness can relate to the struggle of the poor for access to urban resources. When the state and other political institutions categorically deny the poor access to land, the electoral process provides them some hope, which can be seen as an opportunity for candidates willing to take on their case in return for support at the ballot. However, it is often the case that the power relationship between the poor and the patron remains one of extreme inequality, since the former group rarely possesses leadership of political institutions (De Soto, 2000; Bayat, 2004). This opens up lines of enquiry about how the poor are sustained in that unequal power relation. The aim is, therefore, to understand whether – and how – doing nothing can explain the poor's resilience as a form of local urban geopolitical opposition contesting the domination of political elites.

This chapter observes doing nothing as part of the informal nature of urban geopolitical practice – one that reveals micro-tactics and everyday endeavour of the poor in response to electoral politics. While most political science research on South Asia in the 1980s and 1990s has shifted, to a certain extent, towards the analysis of elections and the construction of large-scale models of political behaviour (Piliavsky, 2014), this chapter focuses on the micro-politics of the poor, who are often forced to respond politically to situations within exploitative and disempowering regulations. Such assumptions, however, limit the understanding of the 'technologies of power', whether to a juridico-institutional model of power or a Marxist–Leninist conception of power emanating from 'capital'.[1]

The chapter, however, decisively abandons these models and seeks alternatives by exploring the spatial history of an illegal vegetable marketplace in Khulna, between 1982 and 1990. The vendors within the marketplace managed to occupy a tract of land in *Tarer-pukur* with the help of a patron, after they were forcefully evicted from their original place, *Boro-bazaar*, in 1982. The case of *Tarer-pukur* is explored as part of doctoral research conducted between 2011 and 2016. This particular chapter heavily relies on the interviews, which were conducted in the same period.

I borrow theoretically from Carl Schmitt's concept of *friend and enemy* that delineates why conflict is primordial to order. Schmitt (1976) argues that order arises out of the condition of 'conflict' without fully suppressing it (Neocleous, 1996; Schmitt, 2003). For him, the state makes everything potentially 'conflictual' and political, justified by the recognition of an enemy, which gives legitimacy to its action as a friend of its subjects. Although Schmitt's explanation of friend and enemy suggests why the state as a sovereign occasionally needs to act outside its established order, it also reveals how the state manufactures legitimacy to apply its forces. I use Schmitt's friend and enemy concept to show how doing nothing is a conscious choice in order to make an illegal settlement into a political space.

I argue that doing nothing is a mode of interacting with unequal institutional power[2] by constantly inhabiting the realm between enemy and friend. Doing so creates a balance between institutional power and subsistence. I show that, given the fact that the poor fall outside institutional power structures, inactivity and the construction of an *enemy–friend* relationship create the conflict required to manufacture power inside the illegal settlements, and for survival in an unequal urban geopolitical power-relations.

The eviction and after

In 1982, around 150–200 vendors who regularly traded in the main agricultural marketplace in *Boro-bazaar*, *Bhairab* Strand Road were evicted by the local *arotdars* in negotiation with the state (see Figure 6.1). The *arotdars* were a class of commissioning agents, who sold produce on behalf of the farmers in exchange for money. The vendors, a majority of whom were poor farmers from the countryside, refused to pay the commission to sell their goods inside the *arots* (shops) and were selling on the streets. This deprived the *arotdars* of potential income and, as the popularity of the vendors grew, selling on the street became a threat to the *arotdars*' commission business. Street vendors were subsequently evicted.

Suddenly without a marketplace, the vendors made several attempts to negotiate with the administration and local leaders to obtain another space within the city. However, as per the Bazaar Act of 1951, they were not allowed to sell outside a legally defined marketplace. The 1951 Act aimed to improve the access of poor farmers to markets, and banned private ownership within agricultural marketplaces (East Bengal State Acquisition and Tenancy Act 1950). The local or municipal government, by this Act, became the sole regulatory authority governing marketplaces in urban areas. The vendors therefore required a place approved by the state.

In 1982 (at the time of the eviction), Bangladesh was under martial law, which was later followed by dictatorship (Hyman, 1983). General Hussein Muhammad Ershad usurped state power by removing the constitutionally elected government

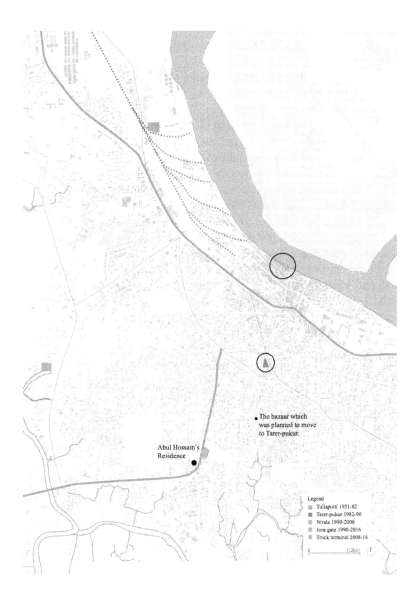

Figure 6.1 Map showing the relocated *Kacha-bazaar* in *Tarer-pukur* (encircled triangle). In 1982, the vendors were evicted from *Bhairab* Strand Road (*Tulla-potti, Boro-bazaar*, encircled).

Source: Author, 2015

of President Abdus Sattar. General Ershad justified the take-over as saving the country from 'economic bankruptcy'. Bangladesh, at that time, was threatened by a grave economic crisis and food shortages, and a power struggle in the ruling nationalist party. Although, throughout the 1970s, the Bangladesh economy experienced acceptable growth rates, this slowed down to an average of 3.1 per cent in the 1980s. The country was granted loans from the World Bank and IMF, but under strict conditions of liberalization and privatization (Rahman, 1992: 98). The decade coincided with the implementation of structural adjustment policies suggested by the World Bank, which meant that the policy reforms executed during this period were not indigenously evolved, but "externally dictated" (ILO–ARTEP, 1993: 205). The fiscal austerity adopted in this period did succeed in bringing down the rate of inflation and reducing the deficit, but caused overall growth rate to fall. Withdrawal of subsidies on agricultural input reduced profit margins, affecting the poor farmers the most (ILO–ARTEP, 1993). General Ershad took over the country's presidency in 1983 and, subsequently, emphasized empowering the local government. Ward elections were, therefore, significant in constituting local government, with commissioners selected from various political parties, based on the residents' votes from the respective wards.

The vendors appealed to various ward commissioners for help to accommodate them in their (commissioners') respective wards. Amid (increased) austerity (for the poor) and liberalization, they frequently heard that there were no places big enough to accommodate them within the town (see Figure 6.2). In reality, the vendors suspected that, since they were poor farmers coming from outside the city to use urban space temporarily, they were not useful to anyone involved in ward-based electoral politics. According to Deloar, one of the destitute vendors, the politicians did not see "much scope (of political gain) in them" (Deloar Interview, August 2012).

After a number of failed attempts, the vendors finally managed to earn some support from Abul Hossain, an aspiring politician, who had influence on a number of political institutions. Hossain was a leader of Jatiyo Party, a former elected commissioner of Ward 24 and the President of Baby-taxi Union in Khulna. Hossain's family also had a long history in politics. Hossain convinced the Municipality Chairman to settle the vendors in *Tarer-pukur*, an abandoned piece of state-owned land in his ward (see Figures 6.2 and 6.3). The land had been prepared to accommodate a bazaar that had never moved there (Hossain Interview, August 2012). The Chairman, however, did not officially approve the settlement, leaving the vendors on a temporary legal footing, so that they could, if necessary, be evicted again (Shohrub Interview, July 2012). Hossain's protection was thus needed to enable them to continue to trade there. The vendors' dependence on Hossain made them a political resource for Hossain, which he (Hossain) could use to his political advantage.

Tarer-pukur *bazaar*

In 1982, the poor vendors started organizing the bazaar in *Tarer-pukur* with a shared spatial culture. There were approximately 164 vendors and the two *chandinas* (shaded structures) with an area of 24 x 96 square feet already built on the site (Figures 6.1 and 6.3). The vendors, however, did not sit inside the *chandinas*;

Figure 6.2 (Disappearing) public land (in grey) inside the city in 2002. Public land, built area within the city and successive locations of *Kacha-bazaar* are shown in grey dots.

Source: Aqua-Sheltech, 2002; Author; and Google Maps

rather they traded their vegetables in open space, similar to their previous way of selling on the street. They used the *chandinas* for storing unsold vegetables at the end of the day so that the rain could not drench them at night (Deloar Interview, August 2012).

Figure 6.3 Tarer-pukur Kacha-bazaar, 1982–90
Source: Author, 2015

The bazaar soon gained popularity in Khulna, as a result both of its low prices and inclusiveness (Rob Interview, January 2014). The temporary and shared culture of space usage accommodated new vendors. In subsequent years, many vendors joined, including Humayun Kabir, who migrated to Khulna from the disaster-prone area of Barisal, near the Bay of Bengal. When he started in 1984, Kabir's financial destitution was apparent, selling vegetables "wearing a torn under-vest and lungi" (Alam Interview, August 2012). Unlike most of the vendors, he was not a producer, but a petty trader who collected vegetables from rural marketplaces, which he sold in *Tarer-pukur*.

The bazaar was managed by Abu-Mia and Kamrul Hasan Badsha, who were also vendors. When the vendors were evicted from *Tulla-potti Kacha-bazaar*,[3] Badshah and Abu-Mia undertook negotiations with influential leaders in Khulna. As a result of their contribution to the resettlement, Badshah and Abu-Mia garnered goodwill among the vendors and were consequently selected as their leaders. In time, Badshah developed a close relationship with Hossain, addressing him as *Abba* (father), and came to act as Hossain's representative in local matters. He took charge of collecting tolls of taka 5 (£0.05) daily from each vendor, to meet the costs of the municipality toll and for running the bazaar. Humayun Kabir did not have any power within the bazaar, nor did the bazaar have any 'real' committee in which he could get involved. In subsequent years, however, Kabir invested a substantial amount of time in dealing with the everyday matters of the bazaar. He took a personal interest in the traders of the bazaar and spent considerable time helping them out, thereby increasing his popularity among the vendors (Kabir Interview, September 2012; Mamun Interview, January 2014). As is seen in other cases of socio-economic exclusion, people tend to use each other as infrastructure – unstable, tentative and temporary, but strong enough to build a degree of economic security (Simone, 2010; McFarlane, 2012; Anjaria and McFarlane, 2011). Kabir's willingness to help and to listen made him a friend of the destitute traders. Often this did not involve him in any activity, just passing time with traders and poor farmers, building relationships. Kabir explained, "this was how people liked me, with my help, people gradually made me their close friend, thinking that this man supports whenever there are problems, so why do we think him as the 'other'" (Kabir Interviews, September 2012 and January 2014). Kabir's time in the bazaar could quite easily be seen as him doing nothing, but the friendships he was forging created status for him within it.

In 1986–7, the vendors organized the election to form a bazaar committee. Kabir was elected as joint secretary with Badshah, while Abu-Mia became its president. Soon after the election, Kabir employed his personal friendship with the fellow leaders to approach them with plans to set up a committee office, offering them the tantalizing possibility of heightened social status among the vendors. He approached Badshah: "*bhai* (brother), let's set up an office, if the office is built then you will get a platform – everyone will call you 'secretary'" (Kabir Interview, September 2012).

Receiving their agreement, Kabir arranged for a collection of funds to build an office with a corrugated iron sheet roof and walls made from bamboo matting. After its establishment, everything was concentrated on that *jhupri ghar* (office). Not only were governing issues discussed there, but it also created a space for the vendors to come and share their everyday problems. For Kabir, this centralization was important because "if it is not centralised, then it is not possible to control" (Kabir Interview, September 2012). The office, like a panopticon, put the vendors in touch with the committee members, but in order to be an omnipresent observer, Kabir needed to spend a lot of time there without doing anything; as he explained, sometimes he spent 'whole nights' in the bazaar, even though selling occurred only in the daytime.

However, as long as Badshah's image as friend was intact, he, not Kabir, maintained real control. Badshah's previous contribution in resettling the vendors meant that he had established trust and agency within the bazaar, his perceived role as a friend of Hossain heightened their faith in him. Kabir objected, however, that while Badshah used to sell potatoes in the previous bazaar (in *Boro-bazaar*) "now he is (simply) 'doing nothing', yet is the key figure in *Tarer-pukur Kacha-bazaar*" (Kabir Interview, September 2012). Kabir believed Badshah's friendship to be double-faced – "*macher teley, mach vejey kheto*" (someone who cooked a fish in its own oil). When a vendor was in trouble, he exploited him to earn money (Kabir Interview, September 2012).

Kabir's position in the bazaar committee was not sufficient to counter Badshah's power – he needed something that could challenge those ties of friendship. Since the friend-enemy is not an absolute distinction, but one that is relative, and each should be seen in the context of a collective, the question for Kabir was how he could make Badshah not merely his own enemy, but a public enemy. Kabir came to hear that Badshah was asking each vendor to donate taka 2,000 (£20) for the development of the bazaar. Badshah proposed to recast all the trading positions in concrete, possibly to protect the mud bases from the rain. There were at least 140 pitches, which meant Badshah could raise a huge sum. As many vendors lived hand to mouth, this led to a dispute. Kabir found that Badshah had pressurized one vendor to pay money for his pitch. Kabir offered full 'moral' support and ultimately extracted a testimony from the vendor accusing Badshah of extortion and of being a threat to the bazaar. Without Badshah's knowledge, Kabir asked the vendors to gather for 'justice' one morning.

Abul Hossian paid regular visits to the bazaar as part of his morning walk. On that day, Kabir invited Hossain for a cup of tea in the committee office. The portrayal of a threat to the bazaar evoked huge interest among the vendors. The committee office was packed with 150–200 people: 'even the *street dog* of the bazaar was sitting in the office!' (Kabir Interview, September 2012).

For Kabir to be able to behead the 'king' amicability was the key stepping-stone, which set the conditions of friendship that justified the creation of an enemy. He stressed Hossain's kindness, underlining that if Hossain took his attention away from the (illegal) bazaar, tomorrow the land would have to be evacuated. The vendors voiced their agreement. A collective recognition of his mercy not only recalled of Hossain's friendship, but also affirmed the vendors as a class of victim. This helped Kabir to play with Hossain's responsibility, which, both knew, was crucial in the political transaction between patron and the vendors.

Kabir continued to argue the importance of Hossain's image and accused Badshah of pressurizing vendors to give taka 2,000 (£20) on the credit gained by using Hossain's name to concrete the bases of their stalls. Kabir alleged that Badshah had forced some vendors to pay the money, thereby abusing their obligation to Hossain. Kabir attacked Hossain with a moral threat to his friendship, arguing that if Hossain demanded the money, then the vendors would offer an apology for whatever they did wrong and then they would go away.

While the vendors' commitment was declared, Hossain's commitment to support the poor vendors was morally challenged. Hossain's patronage operated within the context of a 'transaction' between those who lacked access to resources and someone who held institutional power to provide that access, yet they were dependent on each other for mutual benefit. By targeting his image in front of the poor and presenting his actions as oppressive for the poor, Kabir jeopardized that transaction. The victim politics played by Kabir was powerful enough to portray Hossain as an enemy and to challenge his agency among the vendors.

Badshah was eventually turned into a scapegoat, whose sacrifice was unavoidable for Hossain to save his reputation and retain his dominant position. His institutional power was compromised in the social space of a community of victims, identified by the threat of an enemy. This social space was 'political' not because of the intractability between social and political, but as a result of the orchestration of the enemy, who coerced Hossain into submit to protect his image. Hossain had to play according to the rules set by Kabir, and this resulted in the re-establishing of order by sacrificing Badshah. His institutional power was not of any help in such a social space, as the norms were bent according to the political consent of the vendors. Hossain confessed that Badshah told him that it would be beneficial if the positions of the vendors were given concrete bases. This testimony was sufficient for Kabir to destroy Badshah's authority in the bazaar and his image in front of Abul Hossain.

Badshah's silence and Hossain's vulnerability presented the best opportunity for Kabir to gain authority. By placing a question mark over Hossain's friendship, he established his own agency over the 'order', establishing himself as a superior to Hossain within the bazaar space. This was because the bazaar gave primacy to economic relations related to survival – in which Hossain's image had turned into a liability. Kabir instructed the vendors that if anyone aimed to do something by misusing Hossain's name, then the vendors must first complain to him (Kabir). He insisted that they not only should trade in the bazaar hand-in-hand, but also operate as a clan. The vendors were encouraged to speak to the committee in case of any complaints against anyone, as the committee was there to help and improve matters for them. He asserted that he was "not a 'dictator', but a 'friend'". He also claimed that Hossain was not a dictator – he was also their friend. With Hossain's support for Badshah thus nullified, Kabir became the ultimate friend of the bazaar (Kabir Interview, September 2012; Deloar Interview, August 2012).

The politics of occupation

Unlike many other participants in the bazaar, Kabir was not a peasant but a petty trader who collected vegetables from the rural areas and sold them in the market.

He knew that both farmers and petty traders had an overwhelming hatred of *arotdari* – a form of commission business. This was because a group of commission agents, who were typically known as *arotdars*, had displaced them from their original site in 1982 (Kabir Interview, September 2012). However, in *arotdari*, commission agents had to do little to earn a decent profit. The *arotdars* did not need to produce or collect vegetables from rural areas, but to only occupy a space and rent that to the farmers to sell vegetables. If *arotdars* grew in opposition to the vendors, they would inevitably become a political entity as soon as they had reached a sufficient number.

Kabir picked specific traders among the vendors and passed time with them, frequently making suggestions: "You do *haut-ghat* (petty vending), but take a *ghar*, a fixed place and trade. These [vendors] are all lower-class party, why don't you start the *arotdari* system here? There is profit and honour in it" (Deloar Interview, August 2012). Kabir approached them artfully and sequentially. He convinced them to charge commission from the peasants to sell in their space, that way they could earn as much as they were then earning as vendors. Some traders saw the logic of profit-making and slowly commission business started to grow.

When commission business started in the bazaar, the petty traders who continued to act as vendors began to face losses. They observed that they had to invest labour and pay for transport to collect vegetables from rural areas to sell in the bazaar, whereas the commission agents (*arotdars*) did not have to do anything. Due to the popularity of the bazaar, the farmers were bringing vegetables and were in need of selling space. The *arotdars* could exploit the farmers to give commission for selling vegetables in their occupied space. For the vendors, one option was to continue with the existing system of vending, another was to negotiate with the bazaar committee to secure a space in the bazaar and charge commission from the farmers. This way they could profit by doing nothing.

In subsequent years, in negotiation with the bazaar committee, the number of *arotdars* rose from a few to 20. Abul Hossain was happy with the *arotdari*, as Kabir made each *arotdars* pay Hossain taka 3,000–4,000 (£30–40) monthly for each position (Kabir Interview, September 2012 and January 2014). While *arotdars* were earning money sitting and doing nothing, protecting their commission business from the resistance of the vendors required persistent support from the committee. It is important to underscore that what mattered most was not the sphere within which the opposition (the enemy of the vendors) was unfolding, but the groupings that actually distinguished each other as the enemy – this gave the bazaar committee a continuing political capacity to arbitrate, and a continuing degree of legitimacy.

In an atmosphere of rising tension, the need for the committee to establish order grew ever greater. Kabir and his committee fixed a charge of taka 4 (£0.04) per *maund* (37 kilogrammes) for the *arotdars* to charge from the peasants as commission, so that the vendors did not face substantial losses. For Kabir, the real aim was not to fight according to economic laws, but for political gain. As mentioned previously, the sphere of the political arose from conflict never fully suppressed, which increased both the dependency of the vendors and *arotdars* on the committee.

By constantly being in the realm of friend and enemy, Kabir created power for himself and made the bazaar political as opposed to mere economic space. The committee had to involve Kabir all decision-making – in Kabir's (September 2012) words, "whatever decisions were taken, I would be the final decision-maker". The committee, led by Kabir, took collective action on any issues they considered to be a threat to the bazaar. For example, if someone confronted a trader, the *arotdars*, buyers and labourers all had to gather, as per committee decision, to showcase their power by collective protest. This collectivity was exactly what Hossain needed from the bazaar, although he had to depend on Kabir and the committee to sustain it.

Kabir thus became the vital link between Hossain and the traders of the bazaar. According to many informants, after replacing Badshah, Kabir was often found "sitting together with Hossain all day" – doing nothing (Alam Interview, August 2012; Selim Interview, August 2012). Most importantly, given the political opportunity *Kacha-bazaar* offered through its collective character, Hossain, obliged to secure order in the bazaar, created a permanent authoritarian structure between institutional power and survival of the poor vendors. For example, when President Ershad visited Khulna in 1989, *Kacha-bazaar* traders organized the largest rally in their history to show their support for Abul Hossain (Deloar Interview, August 2012). According to one of the vendors, upon Hossain's request, a meeting was called at which each *arotdar* was instructed to bring 100 people; each vendor had to bring 50 people (Kabir Interview, September 2012; Deloar Interview, August 2012). The *arotdars* and traders knew that failure to comply with the committee's decision could threaten their existence in the bazaar. At that time in *Tarer-pukur* there were at least 40 *arotdars* and 150 vendors, and thousands of people involved in vegetable distribution. At the rally, 55 banners were carried by those from *Tarer-pukur Kacha-bazaar*, with more than 10,000 people in the march (Shohrub Interview, July 2012). The rally was so huge that it completely overshadowed all the rallies of the other politicians. This was surprising to many because Hossain was not an established political leader – merely a member of the ruling party. In the late 1980s, Hossain was elected as a Member of Parliament (MP) for *Batiaghata* – a rural region close to the Khulna and largely dominated by poor vegetable farmers, who also participated in the rally.

The success of the rally and the ensuing election victory not only strengthened Hossain's dependence on the bazaar, but also underscored the importance of doing nothing within the politics of this illegal marketplace. Kabir's actions and motives could be simply reduced to doing nothing, but, by this behaviour, he was able to manipulate the relationship between friend and enemy. However, within a context in which an illegal settlement had no institutional power and required constant patronage, the persistent creation of an enemy generated the capacity for the bazaar to become a political resource – indeed, made the bazaar political. As Jacques Derrida (1993: 355) points out, "losing the enemy", would simply mean "losing the political itself". Drawing on Carl Schmitt, he argues that the 'political', as such, would no longer exist without the figure and without the determined possibility of the enemy. As seen in Kabir's politics, doing nothing created the conditions of necessary conflict within the bazaar and thereby made the space political. Moreover, by making the bazaar space political, he also forged a balance between Hossain's power and the illegal participants.

Conclusion

As I have shown in this chapter, it is significant to understand the internal nature of the political inside an illegal marketplace comprising a group of poor vendors without any institutional political status. It elucidates how the marketplace became more political as the inherent conflict between friend and enemy was created and exploited, and the essential role that doing nothing played in this relationship.

The chapter has elucidated the centrality of conflict in constituting everyday culture and tactics of survival of the poor – the politics of doing nothing. It reminds us that the essence of politics in the illegal settlement is the enemy and doing nothing. In a settlement without legal status, doing nothing, and the production of an enemy, are integral to its sustenance in the face of institutional power. Through the recognition of an enemy, collective claims are tied to moral solidarity. The traders' success in maintaining the occupation of land, as seen in the case in point, depended on their capacity to work as a collective to be useful in a potential political victory of their patron, as well as to articulate the client–patron relationship in their favour.

Further, the role of inactivity in the creation of an enemy–friend relationship reveals something even more decisive about the nature of this secondary city in this decade of structural adjustment and liberalization. *Tarer-pukur*, for example, illustrates how the inclusive nature of the bazaar was gradually replaced by an exclusivist community sentiment through the creation of competing *arotdar* and vendor groups. Although the existence of the groups was necessary for the sustenance of the *illegal* marketplace, those who fell outside the community yet were dependent on it, like the farmers who commuted everyday from rural areas to sell their produce in *Tarer-pukur*, became the new subjects of domination. This indicates a clear trend in the nature of the city. It shows how doing nothing encouraged the formation of a powerful local clique, which transformed the city to a conditional and closed form.

Perhaps more importantly, in this decade of liberalization, the illegal *Kacha-bazaar* in Khulna created a nexus between urban geopolitics and rural manpower. In the closed form of illegal marketplace, doing nothing made spaces like *Tarer-pukur* converging points at which rural poor were integrated with urban politics through their obligation to exchange their produce in urban marketplaces (as seen by the number of the people who gathered in support of Hossain in the rally). Illegal urban marketplaces in this agrarian city acted as the key transactional element in electoral politics, which continued to supply the necessary manpower from rural areas. Breaking the village-centric political structure, they were, thus, internalized within urban-centric geopolitical manoeuvring.

Finally, the internal geopolitics of the bazaar reveals how subaltern vulnerability generates a political resource by the portrayal of an enemy threat, in which doing nothing is an integral tactic. The politics of doing nothing may not fit into the logic of being 'productive' in economic terms, yet it is capable of opening up a different avenue to understanding the ability of the poor slum dwellers negotiating with institutional sources of power. It demonstrates a method of how to better understand the urban poor in the decade of structural adjustment and elucidates their increased effectiveness as a political resource, as well as diminishing their rational choices for defining themselves collectively in order to counteract the pitchfork of liberalization.

Notes

1   Marx believed that the accumulation of capital by the ruling class lies in their interest to secure the ownership of the means of production. The ownership allows the ruling class to control and exploit the working class. See more in Larkin (2011) and Agamben (1998).
2   Here 'institution' refers to political institutions, including political party, labour union, local government and ward commissionership. Institutional power includes party leadership and labour union leadership.
3   In local terms, vegetables market is called *Kacha-bazaar*.

References

Agamben, G. (1998) *Homo Sacer: Sovereign Power and Bare Life*. Stanford, CA: Stanford University Press.

Anjaria, J. S., and McFarlane, C. (2011) *Urban Navigations: Politics, Space and the City in South Asia*. New Delhi and London: Routledge.

Appadurai, A. (1996) *Modernity at Large: Cultural Dimensions of Globalization*. Minneapolis: University of Minnesota Press.

Appadurai, A. (2001) Deep democracy: urban governmentality and the horizon of politics. *Environment and Urbanization*, 13(2): 23–43.

Appadurai, A. (2004) The capacity to aspire: culture and the terms of recognition. In V. Rao and M. Walton (eds), *Culture and Public Action*. Stanford, CA: Stanford University Press.

Aqua-Sheltech (2002) The ownership pattern of public land in the proposed Structure Plan Area. In *Structure Plan for Khulna City 2001–2020*. Khulna: Khulna Development Authority.

Arjun, A. (2004) The capacity to aspire: culture and the terms of recognition. In V. W. M. Rao (ed.), *Culture and Public Action*. Stanford, CA: Stanford University Press: 59–84.

Awan, N., Schneider, T., and Till, J. (2013) *Spatial Agency: Other Ways of Doing Architecture*. London and New York: Routledge.

Bayat, A. (2004) Globalization and the politics of the informal in the Global South. In A. Roy and N. Alsayyad (eds), *Urban Informality: Transnational Perspectives from the Middle East, Latin America, and South Asia*. Lanham, MD: Lexington.

Bayart, J.-F. (2009) *The State in Africa:The Politics of the Belly*, 2nd edn. London: Polity.

Bradshaw, F. (1921) *A Social History of England*. London: University of London.

Chambliss, W. J. (1964) A sociological analysis of the law of vagrancy. *Social Problems*, 12(1): 67–77.

Connell, R. (2005) *Masculinities*. Cambridge, Polity Press.

De Soto, H. (2000) *The Mystery of Capital: Why Capitalism Triumphs in the West and Fails Everywhere Else*. London: Bantam.

Derrida, J. (1993) Politics of friendship. *American Imago*, 50(3): 353–91.

Hyman, A. (1983) Bangladesh under martial law. *Index of Censorship*, 12(4): 4–5.

ILO–ARTEP (1993) Social dimensions and economic reforms in Bangladesh. *Proceedings of the National Tripartite Workshop held in Dhaka*. Bangladesh, 18–20 May: 204–7.

Jeffrey, C. (2010) *Timepass: Youth, Class, and the Politics of Waiting in India*. Stanford, CA: Stanford University Press.

Larkin, C. (2011) An introduction to Michel Foucault's concept of power. *Irish Anarchist Review*, 4: 1–24.

Macfarlane, M. S. P. W. (1961) *Report on the Draft Master Plan for Khulna 1961*. London: Khulna Development Authority.

McFarlane, C. (2012) Rethinking informality: politics, crisis, and the city. *Planning Theory and Practice*, 13(1): 89–108.

Neocleous, M. (1996) Friend or enemy? Reading Schmitt politically. *Radical Philosophy*, 79: 13–23.

Piliavsky, A. (2014) *Patronage as Politics in South Asia*. Cambridge: Cambridge University Press.

Rahman, H. (1992) Structural adjustment and macroeconomic performance in Bangladesh in the 1980s. *Bangladesh Development Studies*, 20(2/3), Trade, macroeconomic policy and adjustment in developing countries (June–September): 89–125.

Roy, A. (2004) The gentleman's city: urban informality in the Calcutta of new communism. In A. Roy and N. Alsayyad (eds), *Urban Informality: Transnational Perspectives from the Middle East, Latin America, and South Asia*. Lanham, MD: Lexington: 147–70.

Roy, A. (2011) Slumdog cities: rethinking subaltern urbanism. *International Journal of Urban and Regional Research*, 35(2): 223–38.

Schmitt, C. (1976) *The Concept of the Political*, trans. G. Schwab. New Brunswick, NJ: Rutgers University Press.

Schmitt, C. (2003) *The Nomos of the Earth in the International Law of the Jus Publicum Europaeum*, trans. L. Ulmen. New York: Telos Press.

Schwandt, T. A. (1989) Recapturing moral discourse in evaluation. *Educational Researcher*, 18: 11–17.

Simone, A. M. (2004) *For The City Yet to Come: Changing African Life in Four Cities*. Durham, NC, and London: Duke University Press.

Simone, A. (2010) *City Life from Jakarta to Dakar: Movements at the Crossroads*. London: Routledge.

Spivak, G. C. (2000) The new subaltern: a silent interview. In V. Chaturvedi (ed.), *Mapping Subaltern Studies and the Postcolonial*. London: Verso.

Townsend, P. (1979) *Poverty in the United Kingdom: A Survey of Household Resources and Standards of Living*. Middlesex: Penguin.

Part III

Urban geopolitics

Middle East and North Africa

Jonathan Rokem

The third part focuses on cities in the Middle East and North Africa, unpacking their profound urban geopolitical transformations. The complex processes of change – from its historical position at the epicentre of three major continents to the recent heightened turmoil infused by the 2011 outbreak of the so-called 'Arab Spring' – moved the region into political instability, civil war and mass migration. At the same time, people in the Middle East and North Africa are undergoing rapid urbanization, economic liberalization and varying degrees of democratization and religious freedom – with these opposing forces making it particularly challenging for cities in this region. Yet the contentious and contradictory processes through which cities are changing remain relatively undefined. Rather, the Middle East has been portrayed all too often in a simplistic manner as swaying between forces of 'anti-globalization' (through Islamic fundamentalism) or 'pro-globalization' (especially in the economic powerhouses of the Gulf city states). Capturing all these complex forces would be impossible within such a brief overview of cases. Yet, in the three chapters, the authors offer a partial – but, at the same time, detailed – window into local urban geopolitical transformations.

Opening the third part, Moriel Ram gives an example from the city of Famagusta, in Northern Cyprus, which has been under Turkish military occupation since 1974, delineating the urban geopolitical legitimacy of Famagusta's urban space by its local residents and nearby university campus as a 'state of exception'. In the second chapter, Nimrod Luz and Nurit Stadler describe the contested urban geopolitics of minority religious claims and religious spatialities and how they are challenged by hegemonic opposing groups in the mixed northern Israeli city of Acre. In the concluding chapter of the third part, Mohamed Saleh writes about Cairo's central position as a pivotal intersection of Africa, Asia and Europe. Saleh, explores the temporal transformation of public space in Egypt and its deep-rooted crisis of participation and identity stretching back from the post-colonial condition to the rise of social media.

7 The camp vs the campus

The geopolitics of urban thresholds in Famagusta, Northern Cyprus

Moriel Ram

Introduction

Drawing from the emergent literature on urban spaces of exception (Boano and Martén, 2013), this chapter critically reflects on the formation process of urban 'spaces of exception' in Famagusta, Northern Cyprus, under Turkish military occupation since 1974. I argue that the conquest and ensuing Turkish occupation constantly produces an urban threshold between Famagusta and several competing spatial processes of encampment, exclusion and seclusion. I will mainly concentrate the reflection around two circumscribed areas situated at the threshold of exclusion/inclusion that constitute and sustain Famagusta. The first is the encircled and enclosed area of Varosha, which represents the overall process of exclusion, seclusion and abandonment that Northern Cyprus underwent following the Turkish conquest while delineating the political existence of Famagusta as a lived *polis*. The second is the campus of the Eastern Mediterranean University, produced as a different form of space of exception directly related to the Turkish occupation, that is insulated from the rest of Famagusta yet substantiates the economic viability of the latter's urban environment. In other words, the urban threshold between Varosha and the city provides the political legitimacy to Famagusta urban space, while the threshold between the campus and the city economically sustain it.

Rethinking the thresholds of the urban encampment

The concept of the camp developed from Giorgio Agamben's seminal work on the state of exception. Drawing from Carl Schmitt's (1988: 5) axiomatic argument that the sovereign "is he who decides on the exception", Agamben argues that the state of exception is based on a paradox: it is promulgated by juridical measures that cannot be understood in legal terms, thus becoming the "legal form of what cannot have legal form" (Agamben, 2005: 1). It is this 'exceptionality' that brings Agamben to deduce that the state of exception can only exist as a liminal threshold between two contingencies: a state of the 'norm' and a state of emergency.

For Agamben, the state of exception is actually a space of exception; an anomic place where, as Schmitt observed, sovereignty becomes pronounced and normal law recedes (ibid.: 31; Schmitt, 1988: 12). The most conspicuous spatial manifestation of the state of exception is the camp where, following Michel

Foucault (1984), Agamben argued that the biopolitics are oriented towards the preservation of the inmate's life on the brink of death (Agamben, 1995: 28–9). The camp can be a Nazi extermination facility (Giaccaria and Minca, 2011), Soviet gulag (O'Neill, 2012), an African centre of humanitarian relief, a site for processing refugees from Middle East (Sigona, 2015), 'villages' for unwanted ethnic minorities such as the Roma in Europe (Picker and Pasquetti, 2015) or underclass population in the USA (Herring and Lutz, 2015). In each of these cases, the camp is a space wherein the Schmittian separation between norm and exception enters a zone of indistinction, where law and violence are rendered indiscernible from one another (Agamben, 2000: 40.1).

Agamben's provocative arguments triggered an intellectual debate around the idea of the camp as ultimate space of exception (Diken and Laustsen, 2005; Minca, 2015; Ek, 2006), which now can be safely regarded as a sub-disciplinary field of camp studies (Minca, 2015). Indeed, the geographies of the camp have enhanced our understanding of the relationship between biopolitics and violence, of management of humanitarian spaces, the analysis of citizenship's actual practices and of individual rights in regimes of humanitarian emergency (ibid.).

Recently, we witnessed a growing interest in the understanding of three main features of the camp's paradigm. First, the proliferation of 'camp studies' suggests the need to revisit the way in which we engage with the notion of the camp itself and its different physical and historical manifestations, by examining how different forms of *encamped* emerge. Encampment alludes to the dynamic in which spaces are shaped into an Agambenian camp through the application of various means of sovereign violence and through practices such as bordering (Sparke, 2006; Amoore, 2006; Epstein, 2007), systematic annihilation (Gregory, 2004), neglect (Shewly, 2013), or by being heir to a broader conflict (Boano and Martén, 2013). Encamped spaces can be formed as interstate frontiers (Hagmann and Korf, 2012), borderlands (Jones, 2009a) and enclaves (Jones, 2009b; Shewly, 2013). In other words, encampment is the process through which a given space becomes a space of exception.

Second, understanding how encampments do not necessarily entail merely the suspension of the norm, but can, at times, produce novel and even transgressive ones in unexpected and unintentional ways (Jones, 2012; Katz, 2015). Stuart Elden (2007) shows how paramilitary training camps can be reshaped into sites of resistance that subvert conceptualizations of territorial sovereignty and integrity. Adam Ramadan demonstrates how Palestinian refugee camps, which have languished in a state of temporariness for over 60 years, have been a frequent target for the exertion of extrajudicial violence (Ramadan, 2009a). These camps, however, also serve as sites for the forging of national identity (Hanafi, 2008) and collective memory (Collins, 2011; Ramadan, 2009b, 2012).

Third, the dynamic through which camps not only suspend the norm but also rearrange it is particularly pertinent in terms of the ways in which urban environment has become the central hub for the formation of the camp localities (Flint, 2009; Schinkel and van den Berg, 2011; Boano and Martén, 2013). As Picker and Pasquetti (2015: 686) point out, there is a need to "bring camps from the periphery to the core of Urban Studies"; encamped urban spaces can become a space of defiance and subversion where norms are challenged and social and political protests unleashed (Ramadan, 2012; Sanyal, 2014; Sigona, 2015).

Considering these elements, this chapter engages with Claudio Minca's recent argument that the current 'camp studies' should examine "what sort of mechanism is in place that allows 'the camp' to be normalized, to operate in some cases just next door to where we live?" (Minca, 2015: 75) – that is, not only to spatially contextualize the form, relations and kinds of violence taking shape inside 'camps' (O'Neill, 2012), but also to juxtapose this contextualization vis-à-vis the surrounding environment. Hence, probing into the urban aspects of the space of exception, and the geopolitics of producing a space of exception in the city, is helpful to understand the convoluted process through which the exception is formed in constant spatial and political dialogue with its surroundings (Herring and Lutz, 2015). As Minca (2015) suggests, not only does the spatiality of the camp determine – in a crucial way – what happens 'inside', but also it contributes to the production of political geographies outside the camp.

In this spirit, this chapter explores the process through which the urban geopolitics of Famagusta is based on delineating a threshold between two encampments, camp and campus, which operate as external borders within the urban environment (Picker and Pasquetti, 2015: 685) and analyses the ways in which they are incorporated into different everyday practices of Famagusta.

Varosha's encampment: separation, suspension and abandonment

Famagusta was Cyprus's main harbour during the medieval period (Önal *et al.*, 1999: 335), the island's capital under Venetian rule (Scott, 2010: 218; Walsh, 2005) and the last outpost that the Ottomans conquered during their invasion of the island in 1571 (Hitchens, 1997). The Ottoman Empire barred the non-Muslim population – the majority of Famagusta's residents – from living within the walls surrounding the city. As a result, most of the city's population moved into an area that was subsequently known as Varosha, a name derived from the Turkish word *varoş* meaning 'suburb' or 'neighbourhood' (Beckingham, 1957). Towards the end of the British rule, in 1958, Famagusta was administratively separated between two municipalities, one per ethnic group (Önal *et al.*, 1999: 336, 339).

Cyprus gained its independence from Britain in 1960. Three years later, the growing animosity between the Greek-Cypriot majority and the Turkish-Cypriot minority led to a protracted inter-communal fighting between the two communities. The period of inter-communal violence was felt in Famagusta, but to a lesser degree than in Nicosia and Kyrenia, the two other major urban localities in what became Northern Cyprus.[1] The reason is, possibly, to be located in the fact that the city was fast becoming a major tourist site, with Varosha as its epicentre – particularly during the late 1960s and early 1970s, when it accounted for about 45 per cent of the entire tourist accommodation on the island (Martin, 1993: 343). At the beginning of the 1970s, Famagusta's population exceeded 40,000 (Keshishian, 1985: 277), Varosha being the most populated district with about 35,000 Greek Cypriots. The Turkish minority was mostly located in the old city.

On 20 July 20 1974, following a coup in Cyprus initiated by the military junta that controlled Greece at the time (Mallinson, 2009: 75–88), Turkey invaded Cyprus. When the attack began, the Greek Cypriot National Guard besieged Famagusta's walled quarter, where most of the Turkish Cypriots hid (Scott, 2002: 220). The Turkish assault on the city was concerned with preventing the Greek Cypriots from attacking the population centre of the Turkish Cypriots. The incursion of the Turkish army commenced on 14 August and was completed within two days, becoming the concluding act of the entire invasion.

From this moment onwards, Varosha became an urban encampment formed through three mutually interacting dynamics. The first was separation. Varosha suffered massive devastation. All of the city's Greek-Cypriot inhabitants (about 35,000), fled or were expelled. Varosha's ethnic cleansing and exclusion mirrored the larger process that took place during the Turkish invasion of 1974, displacing nearly 180,000 Greek Cypriots, whose property was seized. Approximately 60,000 Turkish Cypriots also fled or were driven out of their home towns and villages in the south of Cyprus (Lindley, 2007: 231). The Turkish military captured almost 37 per cent of the island's territory (Kliot and Mansfield, 1997). With a contingency force of around 40,000 Turkish soldiers stationed in the north, the island was cleaved by a 160km Green Line buffer zone controlled by United Nations Peacekeeping Forces in Cyprus (UNFICYP), who had already been active on the island since the 1960s (ibid.: 504). The Green Line marked an extended no man's land with barbed wire, fences, guard posts, walls and barricades (Papadakis, 2005).

A similar dynamic arose in Varosha. Immediately after the fighting and the emptying of the area, the Turkish military moved into the once vibrant and now depleted tourist resort and established a perimeter of about 6 square kilometres around the suburb. None of the civilian population, either Turkish or Greek, was allowed to enter. The military encirclement of Varosha transformed it into a camp that excluded the rest of Famagusta. Those who attempted to get in risked their lives, since the guards were order to open fire on all who tried to enter the new 'camp'. Varosha was formed as an excluded encampment with a process of separation from Famagusta through the ethnic cleansing of the suburb and its seclusion.

In his analysis on the Guantanamo incarceration compound, Simon Reid-Henry (2007) offers a nuanced reading of the politics of the state of exception and the relation between the normal and the abnormal. He argues that, beyond discussing how the camp operates as a spatial anomaly, we need to investigate the internal spatial configuration of the camp and to see how the allocation of space within the abnormal location develops its own mechanisms. Such reading is also relevant to Varosha. A similar principle of separation operates within Varosha's encampment, where some parts of its built environment remained intact and were subsequently transformed into a military barracks for Turkish forces stationed in the area.[2] In addition, Turkish-Cypriot contingency forces were based inside

Varosha under the Turkish military's authority. Nevertheless, the two units are separate and each is allocated a different area within the suburb. The division of the Turkish-Cypriot forces from the Turkish army serves to strengthen the intentions of the latter to present itself as an ostensibly independent polity.

The second aspect of Varosha's encampment is suspension. An exact reason as to why Varosha was enclosed was not given nor was the enclosure presented as an outcome of the fighting. However, it was probably related to the negotiations between Turkish and Greek Cypriots that considered the suburb a valuable asset due to its important tourist industry and its strategic location (Dodd, 2010: 173–6). In July 1978, the Turkish Cypriots offered, following authorization from Turkey, to allow the return of the Greek-Cypriot residents of Varosha and to hand its administration to the United Nations (ibid.: 139–41). As initial contacts between the two sides faltered, the Greek Cypriots did not accept such a proposal. The enclosure was partially explained as Turkish compliance with UN demands, following a plea initiated by the Greek Cypriots to the UN Security Council, which adopted Resolution 550 in 1984, instructing the Turks not to resettle the area (Mayes, 1979). Varosha thus became an asset for the Turkish Cypriots and Turkey, to be used in any negotiation as a potential site that can be returned to Greek-Cypriot control. The transitional nature of Varosha was, somewhat paradoxically, 'fixed', as the area became 'stuck in time' – its final status left 'undecided'. The area was not connected to other parts of Northern Cyprus, which were resettled and became part of the newly shaped Turkish-Cypriot polity. Nor was it given back to the Greek Cypriots and re-established as the island's most attractive tourist hub. Instead, it was placed between the two sides as no man's land (Navaro-Yashin, 2003).

Figure 7.1 The threshold of abandonment between Varosha and Famagusta
Source: Author, 2010

When the fighting concluded, Turkey began a colonization campaign in the territories it had occupied, which included the resettlement of 11,000 Turkish Cypriots (who lived in the south of the island) and about 4,000 Turkish citizens all over Famagusta. This resettlement increased city's population to about 20,000 people (Önal *et al.*, 1999: 340). Among the largest areas repopulated were neighbourhoods adjacent to Varosha, also known as Kato Varosha (Oktay and Conteh, 2007). However, similarly to other resettlement schemes throughout Northern Cyprus (Hatay, 2007), no deeds were given to the new residents (Boğaç, 2009). The lack of ownership placed residents of Kato Varosha in a transient state correlated, to a certain degree, with the experience of exclusion and displacement of the original Greek-Cypriot residents. The latter's ability to be in Varosha was suspended, as they were not allowed to return to the encamped suburb. At the same time, the resettled Turkish-Cypriot presence was constructed as a temporary condition due to the lacks of deeds for the houses they occupied.

The camp is frequently discussed as a temporary site that is "a spatially defined location that exists only for a limited period" (Diken and Laustsen, 2005: 17). In other words, a critical aspect of the encampment relates to the ostensible temporality through which it is facilitated (Belcher *et al.*, 2008).

The enclosure of Varosha constructed it as a space of exception, which is controlled by an outside foreign power (Turkey) – ostensibly, until a final arrangement can be reached (Constantinou, 2008; Kurum, 2012). The 'suspension' of Varosha and its physical effects corresponds with the third aspect of the encampment: abandonment – as a symbol of the Turkish-Cypriot polity. According to Agamben, abandonment correlates with the concept of the ban: an entity is executed by placing the banned individual outside the realm of the *polis*, where political life takes place, and casting him into the wild. Varosha urban space was placed under a ban, dictated by the UN Security Council and enforced by Turkey, as the area's *de facto* sovereign.

Indeed, Varosha's abandonment represents Turkey's exceptional force to exclude whole sections of Northern Cyprus and to render them inaccessible. The massive presence of Turkey's military forces all over the area occupied during the 1974 war was justified as a necessary act in order to protect the lives of Turkish Cypriots, but Turkey did not assume any formal responsibility for the management of the Turkish-Cypriot population. Hence, and in the framework of Northern Cyprus, Turkish forces are placed outside of normative law (Constantinou, 2008) as they are not accountable to local authorities. From this perspective, the ban enforced in Varosha becomes a spatial manifestation of Turkey's sovereign power as it unfolds on the ground: unofficial yet real and concrete.

To put it in Agambenian terms (1998), the birth of Varosha as a camp appeared as an event that decisively signalled the political space of Northern Cyprus itself and is founded on the functional nexus between a determinate localization (Northern Cyprus's territory) and a determinate political-juridical order (the Turkish-Cypriot state). Simply put, Varosha was located outside the new order of the Turkish-Cypriot state and, by that, delimited the newly formed Turkish-Cypriot territory.[3] If, as Claudio Minca (2015: 78) argues, (for Agamben) the concentration camp appears every time we normalize the exception and give it a permanent location, then Varosha's suspension becomes permanent in order to define Famagusta as an urban space. Thus, besides providing a landscape of urban ruination that raises visions of a post-humanist world (Dobraszczyk, 2015), in order to understand how the urban environment of Famagusta is economically sustained we need to move on to another form of encampment, which exists right outside the city's threshold – the campus.

From camp to campus

In November 1983, the Turkish Cypriot area unilaterally declared itself as the Turkish Republic of Northern Cyprus (TRNC). To date, Turkey remains the only state that recognizes the TRNC while the rest of the international community has imposed an embargo that has rendered it completely dependent on Turkey. One successful 'industry' that emerged from this political isolation as a means of compensation was the development of higher education institutions, which have been designed mainly to attract students from Turkey. Since the middle of the 1980s, these universities have become an important income source for Northern Cyprus (Lacher and Kaymak, 2005: 157). The universities produced about $50 million per year revenue to the Turkish-Cypriot economy (Olgun, 1993b: 330). Until the 1990s, five different universities catered to about 8,000 students, 6,000 of whom were from Turkey (ibid.: 327). By 2007, the number of foreign students in these universities amounted to 30,000 (Hatay, 2007: 34) and later increased to 50,000, coming from 68 different nationalities (Güsten, 2014).

The first higher education institution established in Northern Cyprus was the Eastern Mediterranean University, located on the outskirts of Famagusta. Before the 1974 war, the compound where the university stands today operated as a higher education institution of technology. During and immediately after the fighting, the compound was used as a temporary lodging for Turkish-Cypriot families who fled or emigrated from the south (Olgun, 1993b: 330). In 1979, the campus was again converted into a technical institute. It was officially designated as a university in 1986, and its reopening was enabled by funding provided through the Turkish Embassy (İnanç, 2010). By 1990, about 4,600 students enrolled in the Eastern Mediterranean University, mostly from Turkey (Olgun, 1993b: 331). This number increased to 10,000 in 1997 (Önal et al., 1999: 341); currently there are 15,000 students, representing about one third of the total population of Famagusta and its environs, making it the largest university in the north.[4]

When juxtaposed with Famagusta and Varosha, the university's campus can be regarded as an encampment, albeit of a different sort and for different reasons. While it is not enclosed and sealed like Varosha, the campus is distinctively separated from the city's built environment. It is located, like Varosha, on the edge of

Famagusta and not in its centre. However, and unlike Varosha, the borders between the campus and the city are porous. Anybody can go in and out of the campus, yet the campus, nevertheless, is fenced, so admission is only through well-delimited entrances, making the university compound into a site that is separate from the rest of Famagusta.

The higher education institution system is directly supervised by the Turkish Council of Higher Education (YÖK), which sanctions the courses that are taught and reviews the curricula.[5] The majority of the students come from Turkey; a small number originate from the Turkish-Cypriot community, who normally prefer to study in Turkey, where the quality of education is considered to be higher.[6] The YÖK accreditation system allows many Turkish students to attend Northern Cyprus universities if their student scores are too low in the application test for Turkish universities (Güsten, 2014). Thus, higher education becomes a way in which Turkey deepens its involvement, if not control, within Northern Cyprus, embedding itself while maintaining that it is not officially in charge. Hence, like Varosha, the campus becomes a spatial manifestation of Turkey's sovereign hold over Northern Cyprus. Unlike Varosha, it is not accomplished through the preservation of the threat that legitimizes the latter's prolonged presence. The construction of the campus aims to provide Northern Cyprus with a revenue source that will enable its financial independence and, at the same time, link it to Turkey, as the latter provides it with internationally recognized accreditation. Indeed, the opening of the university had a dramatic effect on Famagusta's economy, which reoriented itself towards catering to the students' needs, such as accommodation and day-to-day requirements (Önal *et al.*, 1999: 340; Oktay, 2010).

Thus, the construction of the campus has given Famagusta – seriously affected by the overall conflict – a chance of recuperation. However, from a geopolitical perspective, Famagusta's urban environment is sustained by the campus, which provides the economic lifeline to the city as it suffers frequent economic crises because of its isolation (Bahadi, 2015).

The campus is insulated from the geopolitical conflict. It offers students from other nationalities that suffer from various disadvantages in their own country the opportunity to earn a certified academic degree in more favourable conditions. Mehdi, a student from Iran, told me:

> It is hard to be a secular student in Iran; here, we feel more freedom to do what we want. We have a large community of Iranian students, but we also interact with students from many other countries. We do not constantly feel that we are being watched.[7]

Even more, some feel that the campus offers them a great deal of personal safety. For example, in the 1990s Palestinian students felt that Northern Cyprus's universities were safer and more welcoming than those in the West Bank and Gaza, which were under the direct authority of the Israeli military (Olgun, 1993b: 330). Fodei, a student from Sierra Leone, who spent some time in Saudi Arabia before arriving to Northern Cyprus, provides a different example of the overall feeling of safety:

> When I was in a student in Riyadh, it was customary for the police to stop me for questioning when I came back from visiting friends. I was stopped every five minutes then and asked to show identification documents. I never felt truly safe like here.[8]

This is not to say that the campus of Eastern Mediterranean University is an environment free of threats and a place that offers absolute security. A local Turkish Cypriot, for example, once physically attacked Fodei. Yet, in general, he feels that the campus enabled him a more secure environment than any other place he attended. Furthermore, some of the universities in the north, and particularly the Eastern Mediterranean University, enable students to advance to other higher education institutions in Europe and the USA.[9] Amir, an Iranian PhD candidate in architecture, commented on this subject:

> Students from Iran find it difficult to go directly to universities in Western Europe and Northern America. Because we went to university here, we are considered more 'legitimate'. My wife and I, for example, plan on going to Canada next year. We arrived here, spent the time earning a degree and then we plan to move on to the West.[10]

For some of the students, the campus becomes a place of 'indistinction', of sorts, which is neither located in the Middle East nor exactly in the West.[11]

In the context of Famagusta's overall spatial formation, a remarkable duality surfaces when juxtaposing the encamped Varosha with the university campus. Both camp and campus are located on the rim of Famagusta, geographically and politically. Both operate as sites that were affected by the conflict and the subsequent encampment of Northern Cyprus. Famagusta is thus located between two exceptional sites that exemplify the dual dynamic of encampment and the way it manifests in Northern Cyprus. However, the camp and the campus have an differing effects on Famagusta. The threat Varosha's encampment presents endows the rest of Famagusta political vitality. At the same time, the university campus economically sustains Famagusta.

It is important to notice how Turkey's presence is important in both cases. Turkey's authority over the university is as powerful as it is over Varosha. It provides the academic umbrella that enables the campus to continue functioning and, through this, the sustainability of Famagusta itself. In Varosha's case, it maintains the enclosure of the perimeter and, through that, continues the hazardous situation, which prolongs the conflict. Another common feature of the camp and the campus pertains to the population that resides within each site. Neither the majority of the soldiers stationed at Varosha's military camp nor the majority students are permanent inhabitants of the city or of Northern Cyprus.[12]

Furthermore, both are confined to the city's northern side. Mustafa, a young student from Turkey, shared with me one experience he had. Since he was a relative of a high-ranking Turkish officer, he was allowed to enter Varosha and to tour it from the inside. Yet, when he wanted to visit the south of the island (which has been open to Turkish Cypriots since 2003), he was barred from entering:

> I tried but they [the Greek-Cypriot police] told me that, as a Turkish citizen, I am only allowed to enter Cyprus through the point of entries for international visitors through Larnaka. I was upset because I really wanted to go to the south, I found myself limited.[13]

Residents of both camp and campus are thus confined. Neither can travel to other side of Cyprus. This prohibition is directly linked to the conflict, which produced both camp and campus. Furthermore, the Turkish soldiers and students are somewhat exceptional in Northern Cyprus. Unlike the population that immigrated to the island from Turkey as foreign labour or, originally, as settlers, they are officially excluded from the Turkish-Cypriot body politic. For example, neither of them are included in the official census of the TRNC (Hatay, 2007). At the same time, both the students and the soldiers are connected to sites affecting the geopolitics of Cyprus, and operate as a constant reminder of Turkey's presence – a presence that is, at the same time, invisible and conspicuous, absent and omnipresent, foreign and familiar, a result of an occupation that denies its own existence.

Conclusion

In their analysis of Jerusalem's multiple spaces of exception, Boano and Martén argue that Israel's occupation "consumes Jerusalem… a place where the traditional nation-state is broken and a city of exception is bursting through the cracks" (2013: 10). They note that, in the city, the space of exception is continuously responding, adapting and reorganizing functions of its adjacent environments.

The urban thresholds between Famagusta, the camp and the campus present a different case, where the exception is 'bursting through the cracks' as it permeates into the cityscape. By doing so, the threshold between camp and campus provides a good opportunity to examine the current debate regarding the tension between camps, which epitomizes the notion of a biopolitical machine that spins around an 'empty core', constantly needing to be filled with 'biological substance' (Minca, 2015) and camps that appear as sites where new political subjectivities emerge. In Famagusta, these processes appear side by side, constantly reacting to the city. Varosha has a double, ambivalent effect, contradictory in nature. The suburb's spatial condition serves as a constant urban geopolitical reminder to the national conflict on the island, to the acts of violence wrought by the war of 1974 and to the abnormal presence of the

Turkish occupation. Attempts to normalize Famagusta's space are hampered by the presence of Varosha. At the same time, the separation of the encamped suburb from the rest of Famagusta renders the latter as a place where life is allowed to continue unhindered.

At the same time, the campus, on the other side of the threshold, is isolated – or perhaps insulated – from the geopolitical conflict that purveys Famagusta's city life while it sustains the urban economy and maintains its biological substance by constantly preventing economic collapse. Hence, by understanding the geopolitical positioning of camp and campus vis-à-vis Famagusta, we can provide a partial answer to Claudio Minca's provocative question: "what sort of mechanism is in place that allows 'the camp' to be normalized, to operate in some cases just next door to where we live?" (Minca, 2015: 75). In the case at hand, not only do camp and campus exist 'next door', but they also sustain their neighbour by giving the *polis* that lay beyond it political and economic legitimacy.

Notes

1 One major incident took place on 11 May 1964, when two Greek army officers and the son of the Greek-Cypriot chief of police were shot dead in a gun battle that occurred in the Turkish-Cypriot quarter of the city. In retaliation, the Greek Cypriots carried out an armed sortie that left 35 Turkish Cypriots dead (Patrick, 1976: 68). The incident was severe enough to trigger a formal threat of invasion by the Turkish Foreign Minister at the time (O'Malley and Craig, 2001: 108).
2 Author interview with Mustafa, a Turkish student who visited the complex inside Varosha, 11 March 2013; author interview with C., a Turkish-Cypriot officer who is stationed inside Varosha, 9 March 2013.
3 However, Rebecca Bryant and Mete Hatay's (2011) study of the Turkish-Cypriot enclaves during the 1960s demonstrates that the state of exception and emergency has become pivotal in the formation of the Turkish-Cypriot political legacy prior to the Turkish conquest of 1974.
4 Author's interview with Murat, a representative of the university's registration department, 9 March 2013.
5 Interview with Erol Kaymak and Yucel Vural, Turkish-Cypriot faculty members of the Eastern Mediterranean University, 15 April 2013.
6 Interview with Julian Sarin, faculty member, Middle East Technical University, Northern Cyprus Campus, 5 April 2013.
7 Author's interview, 12 March 2013.
8 Author's interview with Fodei Kunteh, student at the Eastern Mediterranean University, 16 April 2013.
9 Author's interview with Ahmet Sözen, faculty member at the Eastern Mediterranean University, 11 March 2013.
10 Author's interview, 8 March 2013.
11 Interview with Altuğ işığan, Metin Ersoy and Sertaç Sonan, faculty members at the Eastern Mediterranean University, 6 March 2013.
12 There are, of course, exceptions to the exception. During the fieldwork conducted for this research I stayed with a family that had one child stationed as a Turkish-Cypriot officer inside Varosha under the command of the Turkish military and another who was a PhD student at the Eastern Mediterranean University.
13 Author interview, 11 March 2013.

References

Agamben, G. (1995) *Homo Sacer*. Torino: Einaudi.

Agamben, G. (1998) *Homo Sacer: Sovereign Power and Bare Life.* Stanford, CA: Stanford University Press.

Agamben, G. (2000) *Means Without End: Notes on Politics*. Minneapolis: University of Minnesota Press.

Agamben, G. (2005) *State of Exception*. Chicago: University of Chicago Press.

Amoore, L. (2006) Biometric borders: governing mobilities in the war on terror. *Political Geography*, 25(3): 336–51.

Bahadi, N. (2015) Mağusa, Derinya Kapısı'nı bekliyor, *Havadis*, 9 May, www.havadiskibris.com/magusa-derinya-kapisini-bekliyor. Accessed 25 May 201.

Beckingham, C. F. (1957) Islam and the Turkish nationalism in Cyprus. *Welt des Islam*, 5(7): 65–83.

Belcher, O., Martin, L., Secor, A., Simon, S., and Wilson, T. (2008) Everywhere and nowhere: the exception and the topological challenge to geography. *Antipode*, 40(4): 499–503.

Boano, C., and Martén, R. (2013) Agamben's urbanism of exception: Jerusalem's border mechanics and biopolitical strongholds. *Cities*, 34: 6–17.

Boğaç, C. (2009) Architecture for meaning: meaning expression of social value through urban housing in Gazimağusa, North Cyprus, PhD thesis, unpublished document.

Bryant, R., and Hatay, M. (2011) Guns and guitars: simulating sovereignty in a state of siege. *American Ethnologist*, 38(4): 631–49.

Collins, J. (2011) *Global Palestine*. New York: Oxford University Press.

Constantinou, C. (2008) On the Cypriot states of exception. *International Political Sociology*, 2(2): 145–64.

Demetriou, O. (2006) Freedom Square: the unspoken reunification of a divided city. *HAGAR Studies in Culture, Polity and Identities*, 7(1): 55–77.

Diken, B., and Laustsen, C. B. (2005) *The Culture of Exception: Sociology Facing the Camp*. London: Routledge.

Dobraszczyk, P. (2015) Traversing the fantasies of urban destruction: ruin gazing in Varosha. *City*, 19(1): 44–60.

Dodd, C. H. (2010) *The History and Politics of the Cyprus Conflict*. London: Palgrave Macmillan.

Ek, R. (2006) Giorgio Agamben and the spatialities of the camp: an introduction, *Geografiska Annaler. Series B, Human Geography*, 88(4): 363–86.

Elden, S. (2007) Terror and territory. *Antipode*, 39(5): 821–45.

Epstein, C. (2007) Guilty bodies, productive bodies, destructive bodies: crossing the biometric borders. *International Political Sociology*, 1(2): 149–64.

Foucault, M. (1984) Of other spaces, heterotopias. *Architecture, Mouvement, Continuité*, 5: 46–9.

Flint, J. (2009) Cultures, ghettos and camps: sites of exception and antagonism in the city. *Housing Studies*, 24(4): 417–31.

Giaccaria, P., and Minca, C. (2011) Topographies/topologies of the camp: Auschwitz as a spatial threshold. *Political Geography*, 30(1): 3–12.

Gregory, D. (2004) *The Colonial Present*. Malden: Blackwell.

Güsten, S. (2014) Students flock to universities in Northern Cyprus, *New York Times*, 16 February.

Hagmann, T., and Korf, B. (2012) Agamben in the Ogaden: violence and sovereignty in the Ethiopian–Somali frontier. *Political Geography*, 31(4): 205–14.

Hanafi, S. (2008). Palestinian refugee camps in Lebanon: laboratories of state-in-the-making, discipline and Islamist radicalism. In S. Hanafi and R. Lentin (eds), *Thinking Palestine*. London: Zed: 82–100.

Hatay, M. (2007) *Is the Turkish Cypriot Population Shrinking? An Overview of the Ethno-demography in Cyprus in the Light of Preliminary Results of the Turkish Cypriot 2006 Census*. Nicosia: PRIO Cyprus Centre.

Herring, C., and Lutz, M. (2015) The roots and implications of the USA's homeless tent cities. *City*, 19(5): 689–701.

Hitchens, C. (1997) *Hostage to History: Cyprus from the Ottomans to Kissinger*. London: Verso.

Hussain, N. (2007) Beyond norm and exception: Guantanamo. *Critical Inquiry*, 33(4): 735–41.

İnanç, G. (2010) Diplomatic representation of Turkish nationalism and the Cyprus issue. In A. Aktar, N. Kızılyürek and U. Özkırımlı (eds), *Nationalism in the Troubled Triangle*. London: Palgrave Macmillan: 112–30.

Jones, R. (2009a) Agents of exception: border security and the marginalization of Muslims in India. *Environment and Planning D: Society and Space*, 27(5): 879–97.

Jones, R. (2009b) Sovereignty and statelessness in the border enclaves of India and Bangladesh. *Political Geography*, 28(6): 373–81.

Jones, R. (2012) Spaces of refusal: rethinking sovereign power and resistance at the border, *Annals of the Association of American Geographers*, 102(3): 685–99.

Katz, I. (2015) From spaces of thanatopolitics to spaces of natality: a commentary on 'Geographies of the Camp'. *Political Geography*, 49: 84–6.

Keshishian, K. (1985) *Famagusta Town and District. A Survey of Its People and Places from Ancient Times*. Famagusta, Cyprus: Chamber of Commerce and Industry.

Kliot, N., and Mansfeld, Y. (1997) The political landscape of partition: the case of Cyprus. *Political Geography*, 16(6): 495–521.

Korf, B. (2006) Who is the rogue? Discourse, power and spatial politics in Sri Lanka. *Political Geography*, 25(3): 279–97.

Kurum, C. (2012) Arrested development: post-war Famagusta and the distorted perception of its boundaries. Paper presented at the 15th Berlin Roundtable on Transnationality – Borders and Borderlands: Contested Spaces.

Lacher, H., and Kaymak, E. (2005) Transforming identities: beyond the politics of non-settlement in North Cyprus. *Mediterranean Politics*, 10(2): 147–66.

Lindley, D. (1997) UNFICYP and a Cyprus solution: a strategic assessment. Security Studies Program Working Paper: Massachusetts Institute of Technology, Cambridge, MA.

Mallinson, W. (2009) *Cyprus: A Modern History*. London: I. B. Tauris.

Martin, J. (1993) The history and development of tourism. In C. H. Dodd (ed.), *The Political, Social, and Economic Development of Northern Cyprus*. Huntingdon: Eothen Press: 335–73.

Mayes, S. (1979) Can the Cyprus problem be solved? *The Round Table: The Commonwealth Journal of International Affairs*, 69(273): 81–7.

Mbembe, A. (2003) Necropolitics. *Public Culture*, 15(1): 11–40.

McDonald, R. and Makinda, S. M. (1993) Notes of the month. *The World Today*, 49(10): 182–6.

Minca, C. (2015). Counter-camps and other spatialities. *Political Geography*, 49: 90–2.

Navaro-Yashin, Y. (2003) 'Life is dead here': sensing the political in 'no man's land'. *Anthropological Theory*, 3(1): 107–25.

Navaro-Yashin, Y. (2012) *The Make-Believe Space: Affective Geography in a Postwar Polity*, Durham, NC: Duke University Press.

Oktay, D. (2010) Measuring the quality of urban life and neighborhood satisfaction: findings from GaziMagusa (Famagusta) area study. *International Journal of Social Sciences and Humanity Studies*, 2(2): 27–37.

Oktay, D., and Conteh, F. (2007) Towards sustainable urban growth in Famagusta, ENHR2007 Conference Proceedings, Rotterdam.

Olgun, M. E. (1993a) Economic overview. In C. H. Dodd (ed.), *The Political, Social, and Economic Development of Northern Cyprus*. Huntingdon: Eothen Press: 270–99.

Olgun, M. E. (1993b) Sectoral analysis. In C. H. Dodd (ed.), *The Political, Social and Economic Development of Northern Cyprus*. Huntingdon: Eothen Press: 299–335.

O'Malley, B., and Craig, I. (2001) *The Cyprus Conspiracy: America, Espionage, and the Turkish Invasion*. London: I. B. Tauris.

Önal, S., Dağlı, U., and Doraltı, N. (1999) The urban problems of Gazimagusa (Famagusta) and proposals for the future. *Cities*, 16(5): 333–51.

O'Neill, B. (2012) Of camps, gulags and extraordinary renditions: infrastructural violence in Romania. *Ethnography*, 13(4): 466–86.

Papadakis, Y. (2005) *Echoes from the Dead Zone: Across the Cyprus Divide*, New York: I. B. Tauris.

Pasquetti, S. (2015) Negotiating control. *City*, 19(5): 702–13.

Patrick, R. A. (1976) Political Geography and the Cyprus Conflict, 1963–1971, ed. by J. H. Bater and R. Preston. Waterloo, ON: Department of Geography, Faculty of Environmental Studies, University of Waterloo.

Picker, G., and Pasquetti, S. (2015) Durable camps: the state, the urban, the everyday. *City*, 19(5): 681–8.

Ramadan, A. (2009a) Destroying Nahr el-Bared: sovereignty and urbicide in the space of exception. *Political Geography*, 28(3): 153–63.

Ramadan, A. (2009b) A refugee landscape: writing Palestinian nationalisms in Lebanon. *ACME*, 8(1): 69–99.

Ramadan, A. (2012) Spatialising the refugee camp. *Transactions of the Institute of British Geographers*, 38(1): 65–77.

Reid-Henry, S. (2007) Exceptional sovereignty? Guantánamo Bay and the re-colonial present. *Antipode*, 39(4): 627–48.

Sanyal, R. (2014) Urbanizing refuge: interrogating spaces of displacement. *International Journal of Urban and Regional Research*, 38(2): 558–72.

Schinkel, W., and van den Berg, M. (2011) City of exception: the Dutch revanchist city and the urban Homo Sacer. *Antipode*, 43(5): 1911–38.

Schmitt, C. (1988) *Political Theology: Four Chapters on the Concept of Sovereignty*. Cambridge, MA: MIT Press.

Scott, J. (2002) Mapping the past: Turkish Cypriot narratives of time and place in the Canbulat Museum, Northern Cyprus. *History and Anthropology*, 3(3): 217–30.

Scott, J. (2010) Escaping the polarizing gaze: gambling spaces in Cyprus. *Cyprus Review*, 22(2): 291–300.

Shewly, H. J. (2013) Abandoned spaces and bare life in the enclaves of the India–Bangladesh border. *Political Geography*, 32(1): 23–31.

Sigona, N. (2015) Campzenship: reimagining the camp as a social and political space. *Citizenship Studies*, 19(1): 1–15.

Sparke, M. (2006) A neoliberal nexus: economy, security and the biopolitics of citizenship on the border. *Political Geography*, 25(2): 151–80.

Walsh, M. (2005) A gothic masterpiece in the Levant. Saint Nicholas Cathedral, Famagusta, North Cyprus. *Journal of Cultural Heritage*, 6(1): 1–6.

8 Urban planning, religious voices and ethnicity in the contested city of Acre

The Lababidi mosque explored

Nimrod Luz and Nurit Stadler

Introduction

Modern urban planning – characterized by rational, modernistic, centralistic and superimposed approaches to urban design – is becoming increasingly vulnerable (Roy and AlSayyad, 2004; Roy, 2005, 2009; Yiftachel, 2009). Put plainly, the state – through its various agencies – is not the sole player that dictates current developments in the urban landscape. Rapid urban growth, widening diversity, shifting demographics and increasing mobility are leading to the creation of new urban areas and phenomena. These are gradually transforming many cities around the world. In a continuously globalizing world, and with the existence of a growing mixed urban population, planning for *Cosmopolis* becomes a contested task and one that is challenged every day on various levels and by a multiplicity of forces from within and from without the disicpline. The very idea of *Cosmopolis* – becoming the hallmark of postmodernist and critical urban planning under the title of 'celebrating differences' – is impinging on planners worldwide and amounts, on many occasions, to city strife and conflicts (Sandercock, 1998). Various minority groups whose voices were formerly weak or silent in urban politics are starting to build alternative spaces and landscapes, and through them to speak and to emerge as more powerful players (Castells, 1983). These groups are voicing their claims against the power of contemporary nation states that have planned cities based on neoliberal logic, geared primarily to maximizing growth, cost efficiency and accumulation (Harvey, 1989). Religion has been identified as one compelling narrative that serves to mobilize such groups within cities (AlSayyad and Massoumi, 2010; Beaumont and Barker, 2011; Tong and Kong, 2000; Garbin, 2012). Religion provides a useful framework for competing narratives and spatial logic, as well as for the construction of new political geographies in the city. In this chapter we ask: how does religion serve as a driver of urban transformation?

We explore how religious practices, and even buildings, are used by minorities to claim the city and to participate fully in the urban sphere. We show that religion provides a useful framework for the construction of new urban geopolitical geographies in the city. Thus, various distinct groups weave new patterns into the urban landscape through religiously based identity politics (Hervieu-Léger, 2002;

Orsi, 1985). Following previous discussions on the spatial behaviour of religious minorities (Metcalf, 1996; Dodds, 2002, among others), we critically reflect on the construction of religious buildings in urban spaces by minority groups, and their stories, and how they constitute a powerful strategy to reinforce identity and power. Their very presence in the city challenges hegemonic urban forces. Further, religious buildings and discourse provide a focal point for the social integration of minority groups with other groups, and for the recognition of a minority group by the state.

To study the relations between urbanity, religion and ethnicity, in this chapter, we focus on a specific case of the reconstruction of a small mosque – known as the Lababidi mosque, located in the ethnically mixed city of Acre in northern Israel – the struggle to renovate and reopen it after years of disuse, and the debate and conflict that ensued. We examine Lababidi mosque as urban conflict in which the struggle over a religious site and its greater accessibility to a minority's heritage generated a tumultuous and heated debate that divided the city along ethnic and religious fault lines. By focusing on the reconstruction of the Lababidi mosque, we aim to critically reflect the city's transformation in terms of its planning processes, everyday life and politics. We address three questions. How are minority religious claims formulated and given voice through religious spatialities? How are these claims being challenged by hegemonic opposing groups in the city? And in what ways do these voices and actions transform the city's character?

To answer these questions, we followed the conflict and building process of the mosque from 2011 through 2014, and conducted over 20 in-depth interviews (see full interview list at the end of the chapter) with a variety of interlocutors at the city and national levels.[1] The majority of the interviews were with leaders and members of the Muslim minority community of Acre. However, to uncover other voices in the city, we also interviewed members of Acre's Jewish community. In addition, we conducted formal and informal discussions of various issues related to the mosque reinstatement with Acre's municipal officials and with state officials responsible for minority religious compounds. We conducted an ethnographic exploration of the site and thus closely followed the advancement of the mosque reconstruction up until its inauguration. As participant observers, we attended subsequent prayers and gatherings at the mosque. We also investigated the multi-vocality of religious voices that influence, shape and change the city in order to assess the growing importance of religion in urban geopolitics and informality. This investigation was conducted through interviews, as well as surveys of local newspapers and internet material.

The changing urban landscape of Acre

Although the city of Acre boasts a long and meandering history, it remained confined to its historical walls until the beginning of the twentieth century. As the winds of modernization reverberated throughout the Ottoman Empire, changes arrived also to the city in the shape of a modernistic – and, indeed, superimposed – plan that was drawn in 1909 (Waterman, 1971). It left the historic walled old town intact and proposed an orthogonal road system separated by 40m × 50m rectangular blocks for the extramural modern town. During the British Mandate (1918–48), a suburb quickly developed north of the historic walls. It consisted primarily of luxurious townhouses owned by affluent families (Dichter, 1973). In 1930, Ahmad Lababidi, the son of a local wealthy Muslim family, constructed a new mosque as part of a private/family religious endowment (*waqf dhuri*) in the new emerging suburb. This suburb is known, in contemporary Acre, as the Mandatory City. Lababidi mosque was the first mosque to be constructed outside the walls of the historic city and was meant to accommodate the needs of Acre's small population of wealthy Arabs who lived outside the walls (Interview with N, 18 April 2012). A modestly proportioned (20m × 8m) rectangular building was built with fine-cut local stone on an empty lot owned by the Lababidi endowment. The interior was unpretentiously decorated, with the exception of inscriptions of Quranic verses that were finely crafted by a local artist of Persian origin (Interview with M, 27 February 2013). One intriguing feature of the mosque is the absence of a minaret. Our interlocutors suggest that the builder 'omitted' the minaret in an effort to maintain a low profile, given the high socio-economic status and bourgeoisie characteristics of the new neighbourhood.

The establishment of the Jewish state in 1948 had a dramatic effect on the development of the city, on its former Arab urban community and, hence, on the mosque. The Jewish state brought a new understanding and dimension to urban planning. Since its early inception, the Zionist movement adopted a highly modernistic approach to planning and plan implementation as a means of appropriating space (Yiftachel, 2006). The plans aimed to advance the Jewish hegemonic position with regard to space and land (Kimmerling, 1983; Kallus and Low-Yone, 2002). In towns formerly inhabited by Arabs, the new regime implemented a planning policy that aimed to bolster its own dominant position (Yiftachel and Yacobi, 2003). Acre was radically transformed by the tidal waves of political and ideological change following the end of the war. Most obvious of all was the city's demographic shift that left fewer than 3,500 people in the city, a sharp decline from its former population of *c*.15,000 inhabitants. Those who remained were forced to reside within the walls of the historic city (Lurie, 2000). Soon, new arrivals began to alter the urban demography, power structure and urban patterns. The new arrivals consisted mainly of Jewish citizens, especially army veterans and new Jewish immigrants (Kipnis and Schnell, 1978). Initially, these new denizens (or *citadins* as Lefebvre would have had it) lived in vacant houses in the Old City among the Arab residents (Lefebvre, 1991). It would seem that the mixed population of Arabs and Jews in the Old City enjoyed relatively close relationships (Interview with Y, 27 February 2013; Interview with J, 15 August 2011). However, interventions by the planning authorities changed these social dynamics. New neighbourhoods were planned in Acre, earmarked to offer housing solutions for the Jewish population in the newly developed modern parts of the city. The construction of new modern housing outside the historic town – in particular, the planning of the Wolfson neighbourhood in 1963 – sparked the start of a migration that emptied the Old City of nearly all of its Jewish residents (Torstrick, 2000). The vacant lots in the historic city were soon filled with immigrants from Acre's hinterland. These were predominantly marginalized Arab citizens that, for various reasons, decided to change their rural lifestyle and location (Interview with Y, 27 February 2013). The various urban plans implemented in the city followed the ethnocratic logic of separation and reinforcement of the Jewish dominant position. Thus, the Old City was assigned to the Arab population while the newly planned and constructed modern housing was earmarked for the Jewish population (Rubin, 1974). Be that as it may, urban dynamics and demographic changes followed different patterns. Indeed, the Old City came to be predominantly inhabited by Arab citizens. They currently comprise almost 100 per cent of its population (Cohen, 1973; Ben Chetrit, 2011; Interview with Y, February 27, 2013).

As time passed, the new neighbourhood became more and more heterogeneous. By the 1980s, the Wolfson project, formerly a planned Jewish neighbourhood, became the most prominent example of coexistence and cohabitation of Jews and Arabs (Deutsch, 1985; Torstrick, 2000). During the 1970s and 1980s, Acre's Arab citizens bought apartments from Jewish owners. As this trend became even stronger over the years, the cumulative effect of these individual acts was, ultimately, to transform these newly developed urban neighbourhoods.

During this period (1948–90s), the Lababidi mosque shared the same fate as other Muslim religious institutions in the modern parts of Acre; it was closed and banned from use by its former congregation (Interview with M, a Palestinian-Israeli resident, 27 February 2013). Indeed, one must remember that, in the aftermath of the 1948 war, military rule was imposed on the remaining Arab population of Acre until June 1951, and the community lacked the position and resources to challenge the authorities' rulings in any way (Interview with F, 17 January 2013). Legally and officially, the mosque was part of the private Islamic endowment of the Lababidi family, but its ownership rights could not be exercised or even restored (Interview with N, 18 April 2012). Thus, the mosque was left in a state of legal limbo, banned to its former community and unavailable to the local economic forces that could restore it to active use. Local initiatives to restore the mosque to the hands of the community were sporadic and ineffective (Interview with F, 17 January 2013). The Arab minority lacked the relevant skills, political position and public influence to change the course of hegemonic urban planning.

Voicing marginality: the conflict over the Lababidi mosque

In the summer of 2005, the Department for Religious Sects[2] in the Israeli Ministry of Interior initiated a project that provided for limited renovation of the Lababidi mosque. The restoration was confined to the external perimeters of the mosque and was initiated only as a response to a demolition order issued on the compound by the Acre municipal engineer (Interview with Yaacov Salama, 6 December 2013). In light of the state's years of marginalization and disregard of religious minorities' sites, this was a rather unusual move on the part of the Ministry. In order to avoid any misunderstanding or misinterpretation of the goal of the renovation, the head of the department wrote a letter to the director of the al-Jazzar Charitable Trust in which he clarified that the renovation was simply an emergency procedure and that the Muslim community should not regard it as a pretext for any changes in the function (or lack of it as the case may be) or future of the mosque.[3] That is, any future demands or requests by the Muslim community to restore and reopen the mosque for religious activities were still prohibited and would require the state's approval. In Acre, as in other mixed cities in Israel, in the 1950s, shortly after the founding of the state, special committees were established that would officially allow the remains of the Muslim communities to manage their religious sites and communal property. This innovative legal mechanism was initiated and implemented as a means of gaining better control of abandoned or mismanaged Muslim properties (Interview with attorney Victor Herzberg, who represented the Jaffa community in a struggle over the Hassan Bek mosque, 12 July 2002). The committees ostensibly enabled Muslim communities to maintain direct control over their communal and religious property, mostly defined as *Awqaf* (religious endowments). In reality, most committee members lacked the capabilities and were ill equipped to meet the challenges and prowess of state authorities. This was the outcome of screening by the state and various agencies (both in official and non-official capacities) to ensure that highly inept members of the communities, at times with a criminal background or

Figure 8.1 Lababidi mosque
Source: Author 2015

simply illiterate, would be elected to those committees. By and large, they followed their 'master's' voice and did exactly what was expected of them – that is, releasing and leasing their assets cheaply (Interview with Ahmad Asfur, 10 October 2003). In other words, they collaborated with state and municipal authorities and acquiesced to urban planning procedures that necessitated the release of numerous compounds and real estate properties (Luz, 2008).

The emergency renovation of this small building caused a strong phobic backlash among the Jewish citizens of Acre. As they expressed it, their main fear was of an orchestrated campaign launched by the Arab citizens to 'take over the city centre',[4] and the mosque restoration was feared to be part of that.[5] After a post was made on Akkonet, the local internet forum, about the mosque renovation project, blogs appeared expressing grave concern and, basically, suggesting this was just the beginning of a complete and disastrous urban change. An anonymous user, simply named 'aka2005', summed it up as follows:

> This place on Ben Ami Street was never a mosque! It was a Muslim school and not a mosque!!!!!!!!! A mosque is a compound with a dome and a minaret and the roof of this place was always flat. Why do you (i.e. the forum manager) put words in the mouth of the Arabs? And besides, they already dug two holes and casted heavily fortified concrete pillars in order to support the heavy dome to be built here… soon enough, if we will not do something to stop it, this will become an official and sacred place of prayers – to the glory of the Jewish state – the state of Israel.
>
> See Akkonet.co.il (n.d.)

This post reflects the essence of the debate and the almost unanimous views expressed by the participants that the renovation is a very hazardous step towards an imminent Arab 'conquest' of the city centre and the loss of the entire city through this insurgency and de-formalization. In view of subsequent developments, we must emphasize that the Muslim community in Acre had nothing to do with the 2005 renovation.

Given the fact that its leaders were not consulted before or during the renovation, the responses among majority group citizens are rather telling; they mainly narrate, as in the above quoted posts from Akkonet, fear and mistrust in the mixed city. Incidentally, the debate and the emotions it evoked soon subsided, as no visible changes occurred that manifested what was feared – not in the mosque and, certainly, not in the surrounding neighbourhood. There were simply no changes in the built environment that could sustain those phobias and support concrete steps to counter this rudimentary renovation. But this was not the last episode in the convoluted history of the Lababidi mosque.

In 2009, a new director and board members were appointed to the al-Jazzar Charitable Trust. The new board and its director nominated represented a fundamental change, and reflecting the maturation processes Muslim communities have undergone over the years since the declaration of the state in 1948 (Rabinowitz and Abu-Baker, 2005). In Acre, and in the current context of the struggle over the mosque, this change is personified in the newly nominated director of the Trust, Salim Najami.

The son of a prominent local Muslim family, Najami was born and raised in Acre. He serves as a top executive in the regional management of a central bank, and was formerly a member of Acre Unity, an Arab party that was part of the coalition in the Acre municipality (2009–13). He proclaims himself as secular, and does not practice Islam in his daily routine: "I am not a religious man… I came from a place of not knowing what a mosque is, or how many times one has to pray during the day" (Interview, 26 December 2012). It would seem that Najami's attitude and use of religion in the city follows Garbin's (2012: 402) description of urban religiosity: that one "believes in the city" when placing faith in the possibility of sustaining, projecting or even reinventing a sense of self through urban religious place-making and home-making. In this astute observation, Garbin sets the agenda for more inclusive understanding of religious spatial manifestations within the urban built environment. This means to suggest that we need to take seriously the growing influence of religion(s) and religiously motivated groups within cities. This will be readily apparent in what follows, as we discuss the process of reclaiming the Lababidi.

Reclaiming Muslim urbanity: the Lababidi mosque

The Lababidi became part of Najami's agenda following a rigorous 'house cleaning' of the Trust in Acre. Early in 2011, the Trust initiated a complete renovation of the mosque with the unconcealed intention to reopen it and allow it to operate again after a dormancy of 65 years:

> There was a need for a place where Muslims could pray in the new city, indeed there was a need. It is not that I wanted to provoke someone. The mosque was already there in 1948 and I do understand the circumstances [the ample repercussions of the establishment of Israel as a Jewish state], but what I do not understand is the situation in which the majority of the Muslim community lives outside the Old City, but we are not allowed to build anything. We are a third of the city's population. Two-thirds of the community lives outside the walls. Yet they do not have as much as one mosque. Why must an old man who lives in the new city walk and pray in the Old City? I acted with no back-up from any religious authorities and I did not ask for it. I wanted to act quietly and without any publicity or provocation.
>
> Salim Najami, Interview, 26 December 2012

As we reported at length above, Najami voiced his hope for a quiet and non-contentious project. However, this was apparently impossible in the existing socio-political climate of Acre. As soon as the renovation started, the city was abuzz. Again, the local internet forum Akkonet hosted the largely internal debate among its Jewish participants. The reactions ranged from expressions of fear that the city would become dominated by its Arab population (one of the participants exclaimed: "this is a Trojan horse") to accusations of negligence by the mayor for not doing enough to stop this "shameful act". As one of the regular Jewish participants of the forum writes:

> As a citizen that supports coexistence in the city, I respect the freedom of worship of all religions in the city. However, I do not accept a Muslim house of prayer in the middle of the Mandatory City,[6] and the fact is that the city centre is dominated by Jewish citizens.
>
> akkonet.co.il

This citation illustrates the tension between the liberal and postmodern urban planners' attitude towards religious worship and the fear of a Muslim majority. Most responses expressed grave concern for the disruption of the fragile *status quo* between the communities. This is readily apparent in the following excerpt from an interview with Shlomo Fadida, an active and outspoken council member who was in the opposition at that time:

> In principle, I have no problem with the construction of a mosque in Acre. My problem is with its location. It is no secret that there are 'certain' bodies that push Arab residents to purchase flats and assets within Jewish neighbourhoods in order to take them over.... Look at what happened in Wolfson[7]... as soon as Arab families began to buy houses over there, Jewish families began to leave. Next, all the streets in Mandatory Acre that were Jewish are slowly but surely turning into Arab streets. Where will it end?
>
> Akkonet.co.il, accessed March 2012

Najami was aware of these voices and tried to avoid public debates or polemic discussion. However, he remained undeterred even after being approached several times by officials (and others unofficially) from within various corridors of power (Interview, 26 December 2013). On 11 March 2012, the mosque was inaugurated and a prayer service conducted therein – the first since 1948. To avoid further public debate or protests, only 200 local Muslim dignitaries were discretely invited to this festive occasion. Since then, five prayers have been said in the mosque every day, seven days a week. However, the mosque has a few cautious self-imposed restrictions: "The phrase '*Allahu Akbar*' will be said only inside the mosque. We will not say '*Allahu Akbar*' outside the mosque. Perhaps this will happen when the state becomes more democratic" (Akkonet, 22 December 2011).

In this reality, we observe the success of the mosque's role in enabling the infiltration of a religious minority into Acre's landscape and daily existence, providing a prime example of grey spacing and the political geography of informality in Acre. This, we argue, manifests the successful employment of religion by a marginalized minority to change the city and its spaces.

The Muslim voice in the context of Jewish dominance

In her exploration into future cities, Sandercock (1998) is acutely aware of the growing need for a more inclusive, equality-based and democratic city in sharp contrast to monolithic modern-rational urban planning. These ideas, emerging in critical urban theory, were an integral part of implementing demand for the 'rights to the city' as initially suggested by Henri Lefebvre (1991) and later adopted, among others, by Peter Marcuse (2009); they stress the need to restructure the power relations that form the root cause of continuously produced inequality within the urban space (Purcell, 2003). Lefebvre (1996) regarded the right to the city as a cry to struggle to bring forth the necessary shifting and sharing of power from capital and the state to the rightful owners of the city – its inhabitants.

In our analysis, we focus on the revival and resurgence of Muslim religious voices, institutions and faith-based organizations (FBO) in a mixed city, albeit with Jewish dominance, with regard to numbers, politics and control over urban wealth. In accordance with the prevailing ethnocratic logic of urban planning in Acre and the Jewish dominant position in general, the most significant changes are to be found in a variety of Jewish institutions; here we refer to the revival of places such as synagogues, *mikvaot* (ritual baths), *yeshivas* (institutions of higher orthodox Jewish learning), religious schools and so forth. This follows our over-arching argument that the forces shaping the city are more and more frequently emanating from religious voices and demands, mobilized by a variety of religious agents, dominant as well as minority, which collectively contribute to the changing landscape of cities and, surely, the growing influence of religion therein.

In reaction, tensions and rivalries on ethnic and religious grounds are becoming increasingly apparent. This was rather succinctly summed up by Najami, the current head of the Trust, when challenged by a state official who demanded that he close the mosque: "Every back street in Acre has a synagogue, and you can add to this the new compound of the Yeshiva; and what are we asking for is simply to pray!" (Interview, 26 December 2012).

As we observed in our analysis of the city, religious institutions are mushrooming in Acre. In the predominantly Jewish *Shikunei Hamizrach* (Eastern Housing Projects), even a rudimentary survey reveals a surge of religious institutions in recent years. For example, along the main artery of the area one may find a series of religious institutions ranging from a large, ostentatious synagogue to a humble and privately owned daycare centre for toddlers, sponsored by religious FBOs.[8] The inauguration dates posted on some of the buildings' inscriptions indicate their recent construction. The most conspicuous religious compound built to date in this part of the city is the *Ruach Tzfonit* (northern spirit) *Yeshivat Hesder* of Akko. This type of *yeshiva* is a Zionist religious college for students who integrate their army service with religious studies. This organization was established in Acre in 2003 with the express goal of "strengthening the Jewish and Zionist character of the city".[9] The Yeshiva started out in a very humble building in the Wolfson neighbourhood, which, by that time, already had an overwhelming Arab majority. In 2007, ground was broken for a spacious new compound that was later inaugurated on the eastern, Jewish side of the city in a glamorous, well-attended public event held on 30 June 2011.[10] In a circular to his students, Rabbi Yossi Stern, the entrepreneurial head of the Yeshiva,[11] expresses his understanding of the city in theological terms:

> The city of Akko, as the ancient capital of the Galilee and one of the holiest cities in Israel, represents thousands of years of Jewish history. For generations, Akko served as the port of entry to the Holy Land. The Talmud relates that the Sages would kiss the stones of Akko when arriving at this vital seaport. Renowned as a centre of religious study, Akko attracted many great scholars including the Ramhal (Rabbi Moshe Chaim Luzzatto), Maimonides and Nachmanides. Among the famous visitors to Akko's Jewish community was Benjamin of Tudela, the Jewish traveler and diarist. Akko is remembered as the site of the 1947 prison breakout by Etzel [a Jewish paramilitary organization]. Later, the city became home to many Holocaust survivors seeking refuge at Israel's shores.[12]

In this letter, Stern promotes and enhances the Jewish narrative of the places/city by assigning famous Jewish scholars such as Ramhal, Maimonides and Nachmanides to Acre. Through these historical anecdotes, Stern reconstructs a Jewish chain of memories (Hervieu-Léger, 2002). This establishes the Yeshiva as an important link in an already well-founded Jewish heritage (and claims) in the city. The growing influence of religious voices is very obvious also among the Arab (mostly Muslim) community of Acre. A few Islamic FBOs operate in the city in addition to the al-Jazzar mosque described above and the newly renovated Lababidi mosque. In the early 2000s, a local section of the Israeli Islamic Movement, based at Masjid al-Ramal (literally, the sand mosque) at the centre of the Old City, was responsible for affixing signs that promoted Islamic piety and repentance. These trends towards claiming the rights to the city are part of a process of the construction of a counter-hegemonic political urban geography. Further, they also form one element of an ongoing struggle over symbols, materials and meanings in the urban landscape, a struggle that has, in recent years, been characterized by ceaseless efforts on the part of Acre's Arab citizens to counter and challenge the superimposition of street naming by the state (Shoval, 2013).

The green signs (the colour of the Islamic Movement)[13] and their content fit well with the organization's general attitude of challenging the Jewish character of the state of Israel (Luz, 2013). Their very presence in the urban landscape, especially when posted in juxtaposition to the official ones of the state and municipal authorities, are a very clear counter-hegemonic statement. Indeed, they lay claim to better representation and participation in the urban landscape. The Islamic Movement is also responsible for a series of renovation projects in various Islamic compounds in the Acre Old City.[14] These projects focused mostly on renovating religious compounds such as a dilapidated mosque and a cemetery surrounding a local shrine. We argue that these actions are possible because of the growing strength of the informal forces described above. Muslim actors use these spaces to reinforce their religious identity, and mosque-building is a primary and effective strategy within these local urban geopolitical conflicts.

However, as we previously suggested, this state of affairs does not guarantee better representation of citizens in planning procedures or in city policies, nor does it naturally transform the city into a more inclusive one. These institutions and their compounds challenge contemporary planning and regulations in the city and allow for marginal groups to claim the city through those challenges on the prevailing spatial hegemonic logic, thereby creating new geopolitical urban geographies. The urban transformations of space and society in Acre indicate the growing importance of religion and religious voices as a frame of reference for various claims of the city.

Conclusion

This chapter started by questioning the changing nature of urban landscapes and everyday life through religion and religious infrastructures. We asked how religion serves as a platform of urban transformation. Our main argument is that people who find themselves deprived, muted and disposed of are now more frequently turning to religion, religious buildings and religious voices as mechanisms by which they establish and maintain their presence in the city. Acre, as a case in point, suggests that the use of religious institutions is a central part in an urban geopolitical ethno-national struggle to manifest the minority community's presence in space as part of their claim for a right to the city. We analysed the case of the Lababidi mosque in Acre to investigate these changes. We demonstrated that urban spaces and landscapes in the city of Acre are becoming more strongly influenced by religious claims and religious buildings. These struggles over religious sites should be seen as a mechanism that marginalized urban citizens can use to challenge conventional property rights and hegemonic domination of the urban landscape (Mitchell, 2003; Purcell, 2003; Yiftachel, 2009; Shoval, 2013).

Our case in point – that is, the renovation of the Lababidi mosque – illustrates how religion has, indeed, become a compelling narrative that serves to mobilize minority and subaltern groups within cities (AlSayyad and Massoumi, 2010; Beaumont and Barker, 2011; Tong and Kong, 2000; Garbin, 2012). The struggle over the mosque and the role played by Salim Najami, the chair of the al-Jazzar Charitable Trust, indeed corroborate contemporary views among scholars of urban religiosity and demonstrate how religion provides a useful framework for competing narratives and spatial logic, as well as for the construction of new geopolitical geographies in the city (Hervieu-Léger, 2002; Orsi, 1985). Our arguments rely also on the premise that the claiming of religious heritage by religious minorities (as it is transformed into the physical form of religious buildings) not only informs their religious identity, but is also closely associated with emancipatory and highly politicized processes. These processes render minorities and marginalized groups more susceptible to scrutiny and suspicion as they intensify their contest with hegemonic discourse and construct visible signs (i.e. religious sites) that rise from religious alterity, different spatial logic and different understanding of the city. Thus, echoing Orsi (1999), conflicting voices, visions and desires emerge from religiously framed identity politics to contend with and prevail against modernistic planning procedure and spatial logic.

Notes

1 Interviews ordered according to date: Interview with attorney Victor Herzberg, 12 July 2002, his office in Tel Aviv; Interview with Ahmad Asfur, activist in the Hassan Bey struggle, 10 October 2003, local resident in Jaffa; Interview with J, resident of Old Acre, 15 August 2011, al-Ramal Mosque in the Old City of Acre; Interview with M (of the Lababidi congregation), independent, 18 April 2012, coffee place at the Old City; Interview with Mr Shimon Lankri, Mayor of Acre, 20 April 2012, Acre municipality; Interview with A, retired NGO worker, 18 April 2012, local resident in Acre; Interview with Salim Najami, Director of the al-Jazzar Charitable Trust, 26 December 2012, coffee place in Acre; Interview with Imam Samir Asi, 5 January 2013, al-Jazzar Mosque; Interview with D (of the Lababidi congregation), local business owner, 11 January 2013, local resident in Acre; Interview with S, local businessman, 15 January 2013, his workplace; Interview with F (of the Lababidi congregation), retired clerk, 17 January 2013, local resident in Acre; Interview with M (of the Lababidi congregation), a retired teacher, 8 March 2013, local resident in Acre; Interview with A, student-activist, 24 February 2013, coffee place in Acre; Interview with H, student-activist, 24 February 2013, coffee place in Acre; Interview with M, resident in the Old City, 27 February 2013, local resident in Acre; Interview with R, resident in the Old City, 27 February 2013, local resident in Acre; Interview with Shaykh Muhammad Zahra, Imam of the Lababidi mosque, 27 February 2013, Lababidi mosque; Interview with B, student, March 12, 2013. At the Western Galilee College; Interview with L. (a student), March 12, 2013. At the Western Galilee College; Interview (by phone) Yaacov Salama, Head of Religious Sects, Ministry of Interior, 6 December 2013; Interview with R, resident of Wolfsson neighbourhood, 8 April 2014, local resident in Acre; Interview with Y, retired head of community centre and activist, 24 April 2014, local resident in Acre; Interview with A, student, 21 May 2014, Western Galilee College; Interview with Rabbi Shmuel Yashar, 25 September 2016, his office in Acre.
2 I.e. non-Jewish religious groups in Israel.
3 A letter sent by Mr Yacov Salaama, Head of the non-Jewish Religious Groups division at the Minstry of the Interior, to Mr Kamal Jayyushi, director of the al-Jazzar Charitable Trust, 23 November 2004.
4 This is a common phrasing, often narrated by our Jewish interlocutors in the city.
5 Data collected mostly from a local urban internet forum, www.akkonet.co.il. Accessed 23 April 2016.
6 This is the part of the city that was developed mainly during the Mandate period, 1918–48.
7 As mentioned above, Wolfson is the name of the neighbourhood that was built in the 1960s and was earmarked for Jewish citizens. It eventually became an ethnically mixed neighbourhood.
8 This is the finding of an ongoing survey of the city conducted by one of the authors. The survey is part of a study of religious institutions in Acre conducted by Luz since 2010.
9 See http://yakko.co.il/eng/. Accessed 23 April 2016.
10 See www.akkonet.co.il/forums/viewtopic.php?f=8&t=66. Accessed 23 April 2016.
11 See the letter: http://yakko.co.il/eng/maamar.asp?id=33661&cat=2577. Accessed 23 April 2016.

12 Stern, http://yakko.co.il/eng/maamar.asp?id=33661&cat=2577. Accessed 23 April 2016.
13 Since the early 1970s, an Islamic movement that follows the creed of the Egyptian Muslim Brothers operates in Israel. As with many others branches of this ideological movement, its emblem is a green flag with Quranic inscriptions that adorn it.
14 www.panet.co.il/online/articles/1/2/S-316459,1,2.html. Accessed 23 April 2016.

References

Akkonet.co.il (n.d.) www.akkonet.co.il/forums/viewtopic.php?t=294. Accessed June 2016.

AlSayyad, N., and Massoumi, M. (eds) (2010) *The Fundamentalist City? Religiosity and the Remaking of Urban Space*. London: Routledge.

Beaumont, J., and Barker, C. (eds) (2011) *Post-Secular Cities: Space, Theory and Practice*. London and New York: Continuum.

Ben Chetrit, G. (2011) *Urban Renewal in Old Akko, Local Residents: A Burden or a Boon.* Tel Aviv and Haifa: Tel Aviv University and BRM Institute of Technology and Society.

Castells, M. (1983) *The City and the Grassroots: A Cross-Cultural Theory of Urban Social Movements*. Berkeley: University of California Press.

Cohen, E. (1973) *Integration vs Separation in the Planning of a Mixed Jewish–Arab City in Israel*. Jerusalem: Levi Eshkol Institute for Economic, Social and Political Research, Hebrew University of Jerusalem.

Coleman, S., and Maier, K. (2013) Redeeming the city: creating and traversing 'London–Lagos'. *Religion*, 43(3): 353–64.

Deutsch, A. (1985) Social contacts and social relationships between Jews and Arabs living in a mixed neighborhood in an Israeli town. *International Journal of Comparative Sociology*, 26(3–4): 220–25.

Dichter, B. (1973) *The Maps of Acre: An Historical Cartography*. Acre: Municipality of Acre.

Dodds, J. (2002) *The Mosques of New York City*. New York: Power House.

Garbin, D. (2012). Introduction: believing in the city. *Culture and Religion*, 13(4): 401–4.

Harvey, P. (1989) *The Condition of Postmodernity: An Enquiry into the Origin of Cultural Changes*. Oxford: Blackwell.

Hervieu-Léger, D. (2002) Space and religion: new approaches to religious spatiality in modernity. *International Journal of Urban and Regional Research*, 26(1): 99–105.

Kallus, R., and Law Yone, H. (2002) National home/personal home: public housing and the shaping of national space in Israel. *European Planning Studies*, 10(6): 765–79.

Kimmerling, B. (1983) *Zionism and Territory: The Socioterritorial Dimensions of Zionist Politics*. Berkeley: University of California Press.

Kipnis, B. A., and Schnell, I. (1978) Changes in the distribution of Arabs in mixed Jewish–Arab cities in Israel. *Economic Geography*, 54(1): 168–80.

Lefebvre, H. (1991) *The Production of Space*. Oxford: Blackwell.

Lefebvre, H. (1996) *Writings on the City*. New York: Wiley-Blackwell.

Lurie, Y. (2000) *Acre City of Walls: Jews Among Arabs, Arabs Among Jews*. Tel Aviv: Yaron Golan Press.

Luz, N. (2008) The politics of sacred places: Palestinian identity, collective memory, and resistance in the Hassan Bek mosque conflict. *Environment and Planning D: Society and Space*, 26(6): 1036–52.

Luz, N. (2013) The Islamic movement and the seduction of sanctified landscapes: using sacred places to conduct the struggle for land. In E. Rekhess and A. Rudnitzky (eds),

Muslim Minorities in Non-Muslim Majority Countries: The Islamic Movement in Israel as a Test Case. Tel Aviv: Eyal Press, Tel Aviv University, Konrad Adenauer Program for Jewish–Arab Cooperation: 67–78.

Marcuse, P. (2009) From critical urban theory to the right to the city. *City*, 13(2–3): 185–97.

Metcalf, B. (1996) Introduction: sacred words, sanctioned practice, new communities. In B. Metcalf (ed.), *Making Muslim Space in North America and Europe.* Berkeley: University of California Press: 1–27.

Mitchell, D. (2003) *The Right to the City: Social Justice and the Fight for Public Space.* New York and London: Guilford Press.

Orsi, R. A. (1985) *The Madonna of 115th Street: Faith and Community in Italian Harlem 1880–1950.* New Haven: Yale University Press.

Orsi, R. A. (1999) *Gods of the City: Religion and the American Urban Landscape.* Bloomington: Indiana University Press.

Purcell, M. (2003) Citizenship and the right to the global city: Reimagining the capitalist world order. *International Journal of Urban and Regional Research*, 27(3): 564–590.

Rabinowitz, D., and Abu-Baker, K. (2005) *Coffins on Our Shoulders: The Experience of the Palestinian Citizens of Israel.* Berkeley: University of California Press.

Roy, A. (2005) Urban informality: toward an epistemology of planning. *Journal of the American Planning Association*, 71(2): 147–56.

Roy, A. (2009) Strangely familiar: planning and the worlds of insurgence and informality. *Planning Theory*, 8(1): 7–12.

Roy, A., and AlSayyad, N. eds. (2004) *Urban Informality: Transnational Perspectives from the Middle East, South Asia and Latin America.* Lanham: Lexington.

Rubin, M. (1974) *The Walls of Acre: Intergroup Relations and Urban Development in Israel.* New York: Holt, Rinehart and Winston.

Sandercock, L. (1998) *Towards Cosmopolis: Planning for Multicultural Cities.* London: John Wiley.

Shoval, N. (2013) Street-naming, tourism development and cultural conflict: the case of the Old City of Acre/Akko/Akka. *Transactions of the Institute of British Geographers*, 38: 212–26.

Tong, C. K., and Kong, L. (2000) Religion and modernity: ritual transformations and the reconstruction of space and time. *Social and Cultural Geography*, 1(1): 29–44.

Torstrick, R. K. (2000) *The Limits of Coexistence: Identity Politics in Israel.* Ann Arbor: University of Michigan Press.

Waterman, S. (1971) Pre-Israeli planning in Palestine: the example of Acre. *The Town Planning Review*, 42: 85–99.

Yacobi, H. (2003) The architecture of ethnic logic: exploring the meaning of the built environment in the 'mixed' city of Lod – Israel. *Geografiska Annaler*, 84(3–4): 171–87.

Yiftachel, O. (2006) *Ethnocracy: Land and Identity Politics in Israel/Palestine.* Philadelphia: University of Pennsylvania Press.

Yiftachel, O. (2009) Critical theory and 'gray space': mobilization of the colonized. *City*, 13(2–3): 240–56.

Yiftachel, O., and Yacobi, H. (2003) Urban ethnocarcy: ethnicity and the production of space in an Israeli mixed city. *Environment and Planning D: Society and Space*, 21: 673–93.

9 Exploring the roots of contested public spaces of Cairo

Theorizing structural shifts and increased complexity

Mohamed Saleh

Introduction

The condition of urban public spaces in Egypt has been one of the major drivers of the massive social upheavals that the country witnessed from 2011. Prior to these events, what was nominally called 'public space' has never been much public. Its *publicness* has been discovered or rather seized – although temporarily – by the popular majority during the upheavals. Being public means that it functions to serve the different, and often agonistic public interests, *all* of them. That is, of course, if they were counted as legitimate parts of the 'accepted' public from the first place. Perhaps this may sound too idealistic, but, in fact, this is the heart of the problem. The right to public space was inclusive to powerful political, economic and security agents who constitute the *police order* and the rest was considered mere *noise*, to use Rancière's word (1999).

Starting from the adoption of neoliberalism in the early 1970s, the state has conceived all urban public assets as objects of economic development and tools to consolidate power, especially public spaces. In his critique of neoliberalism, Simon Springer (2010: 455) accurately describes this as the "neoliberal assault on all things public" – that is, the attempt of the state to dictate an 'ordered' public space. This goes against the force of ordinary citizens demanding a free and 'unscripted' public space (Springer, 2010). Within such space, they can express their different identities and feel as active participants, whose agency is recognised in the *police order*. Rancière portrays that order as a realm designed by the state which curtails the voices or identities allowed to be seen and heard in public: an act he defines as *partition of the sensible* (Rancière, 1999). Tensions therefore arose between those two opposing forces, and public space became increasingly contested over the years.

Addressing this contestation in Egypt, I incorporate notions derived from complexity science to explore the roots of contestation. This approach forms an anti-reductionist framework for challenging some mainstream labels, which describe the public sphere of the Arab-Muslim region (Ben Moussa, 2013). For example, the label of 'Arab street' implicitly refers to a backward and apathetic place full of angry mobs (Bayat, 2010). The lack of contextualized understandings is another source of reductionism – that is, the tendency to squeeze the Arab issues into Euro-centric debates, which cannot account for the socio-historical layers of the context. By contrast, a relational approach of complexity allows for deriving a more contextualized and historically relevant understanding. Furthermore, the Egyptian 2011 upheavals were part of a broader emancipatory geopolitical context. The synchronized affinity between various contestations in the region led to over-emphasizing the commonalities. While important commonalities cannot be denied, theorizing the historical specificities of Egypt can explore more missing pieces of the puzzle in understanding these events. Thus, in this chapter, I set out to explore the roots of the contested public spaces of urban Egypt, tracing their genealogical development on the *longue durée*. In other words, I attempt to answer the question: what are the structural conditions that facilitated the emergence of this urban geopolitical contestation on an extended timeline?

The argument in this chapter is parcelled into two main parts. First, I briefly discuss four key principles of complexity science. Then, drawing on those principles, I critically advance the complexity perspective as an alternative approach to address the urban contestation of the Arab context. By giving illustrative examples from the Egyptian case, I begin to bridge the general principles of complexity with a real-life phenomenon. Second, an overview of the Egyptian socio-political landscape is provided, focusing on its influence on urban public spaces. Then, I employ the principles of complexity to analyse the case. In particular, the principle of path-dependency is used to scrutinize the socio-historical development of the contested public spaces. Aiming to explore the roots of such contestation, I focus on four historical thresholds at which the structural condition of public space changed.

Reduced complexity and deterministic labels

Depicting the contestation of contemporary public spaces as a complex phenomenon means embracing an anti-reductionist and relational approach. Complexity science offers particular mechanisms of understanding that challenge much of the reductionism that obstructs forming a contextualized and historically relevant understanding of the roots of contestation. However, I stress that complexity science should not be perceived as an all-encompassing approach to understand reality. In an abstract sense, complexity is an alternative lens to view particular conditions of reality, which comes with its own principles of conceptualization. Of relevance to this chapter are four of those principles: (1) the rejection of reductionism and determinism; (2) dynamic systems can shift structurally between different degrees of complexity; (3) change in complex systems occurs in abrupt leaps between order and chaos; and (4) change in complex systems is mostly path-dependent.

Principles of understanding

First, complexity science is essentially against ideas that relate to the so-called 'Cartesian reductionism'. This originated from Rene Descartes's philosophy.

Descartes based his methods on understanding the human biological parts and functions as machines (De Roo, 2012). Dissecting of the separate parts and analysis of their functions by expert scientists led to an accurate knowledge of the machine as a whole. Ravetz and Funtowicz (1993) categorized this as a 'normal' scientific mind-set as opposite to the 'post-normal' mind-set, with the latter being a more suitable to understand and respond to complexity. They argue that the normal scientific mind-set traditionally peruses a regulated, simplistic and predictable picture of reality: a world in a durable equilibrium (De Roo, 2012). This can be attainable in straightforward conditions, but would be highly irrelevant in complex conditions. For example, the dominant spatial logic of neoliberal politics often strives for an *ordered* and harmonized spectacle of public spaces. However, public space is a complex phenomenon that has the potential to promote democracy through a dynamic agonistic dialogue between diverse political subjects and groups (Springer, 2010; Mouffe, 2006). Thus, by seeking to establish order, neoliberalism is reducing the full potential of public spaces.

The second principle classifies different degrees of complexity. By virtue of being highly dynamic, complex systems are in a discontinuous back-and-forward motion between simple and complex situations (De Roo, 2012). This goes against Kuhn's theory of conceptualizing reality as "an unquestioned and unquestionable paradigm" (Kuhn, cited in Ravetz and Funtowicz, 1993: 5). Scholars who bridge complexity science with spatial planning emphasize that development may occur due to structural shifts in the context in which the system is embedded (De Roo, 2012; Portugali, 2006). Most common of such shifts is from a closed system to an open system. A system is said to obtain a 'higher degree' of complexity when the surrounding context experiences such a shift. Higher complexity means entering a highly uncertain, unstable and fuzzy context.

Instability and fuzziness thrive because open systems intensify the traffic of information and energy between the parts, making the system seemingly move towards chaos (De Roo, 2012). For instance, contemporary public spaces are, in a way, witnessing a similar shift. The 'ordering' force of neoliberalism has bound the material space in a closed system for decades, denying a truly dynamic public space of agonistic politics (Mouffe, 2006). But, due to the massive spread of social media, the ordered material space is interdependently coupled with dynamic virtual spaces (Castells, 2015). The agonistic feature of a democratic public space has found alternative pockets of freedom. Moreover, social media increases the attributes that typify complex systems such as connectedness, adaptiveness and diversity (Tierney, 2013). Thus, by becoming coupled with the virtual spaces of social media platforms, I shall argue in this chapter that public space structurally shifted from a closed to an open system.

Third is the principle of bifurcations, where change in complex systems does not occur as a smooth development due to linear cause and effect. It takes the shape of sudden abrupt events. Using the vocabulary of complexity science, these are the *bifurcation* events. At a bifurcation, the system may jump unexpectedly from a stable to chaotic state (Portugali, 2006). In a socio-political sense, a bifurcation might manifest itself as a major technological shift, an economic crisis or a 'revolution'. However, complex systems never scatter into total chaos, but they rather maintain themselves on the 'edge of order and chaos'. A complex system such as a living organism is either 'moving towards equilibrium or away from it', continuously inhibiting this dynamic in-between (De Roo, 2012).

But what makes complex systems able to shift between order and chaos? This is largely due to the capacity of *learning* that typifies complex systems. If a complex system is being forced to stay in a stable state (no matter for how long), we should never overlook the destabilizing capacity of its sub-systems. Even if a complex system appears stable on the surface, the adaptiveness of its parts enables them to transform themselves by continuous learning in response to contextual changes. During the bifurcation phase this learning capacity crystallizes as the parts co-evolve with the context to a higher level of complexity. For example, the hegemonic desire of the Egyptian state to de-politicize citizens locked them out of official institutions of representation, pushing them to find alternative spaces to exercise their human need of becoming part of the public (Kattago, 2012). Shenker (2016: 264) eloquently describes these spaces as "imaginative pockets of Egypt in which life and politics play out in a completely different way". The citizens' response reflects their learning capacity and ability to govern their own lives away from state eyes.

Fourth comes the principle of path-dependency; change in complex systems does not evolve randomly but due to preceding events that condition the route of development. History, remarkably, matters. Path-dependency means that the first few choices (thresholds) along the path strengthen the chance of certain developments to actualize, and hinder chances for other possibilities (it is not the initial state alone that matters). The path is *self-reinforcing* itself. However, in complex systems there are always a vast number of possibilities that the system might strengthen or hinder. One cannot decide *a priori* which future will take place. Also, every threshold along the path creates unexpected outcomes. The path is therefore *non-linear*. To conclude, this principle entails that change takes the shape of "non-linear *self-reinforcing* processes" (Ebbinghaus, 2009: 191, emphasis added).

Cairo's public spaces provide an appropriate example of such principles. As will be discussed later in the chapter, the Egyptian state has followed deliberate policies and physical interventions to render public space devoid of any political meanings: it *de-politicizes* space. But the state does not have enough manpower and institutional capacity to enact this desire upon the totality of Cairo's public spaces. Instead, it has focused its efforts on epicentres of the city such as Tahrir Square and Ramses Square. Due to the highly centralized nature of Egypt, not only Cairenes but also every Egyptian has to pass by those squares at least once in their lifetime. Subsequent interventions towards de-politicizing such spaces have enabled the state to set the dominant trend. The trend has been imprinted in the mental map of the users of space and reinforced itself by being implemented in other minor spaces, without the presence of the state. Thus, more de-politicizing begets more de-politicizing, at the same time oppressing alternative versions to the dominant trend.

The first three principles guide how the issue of contested space is being conceptualized in this chapter. To operationalize the complexity thinking around the case study, the fourth principle of path-dependency is utilized to explore the very roots of contestation. Empirically, the chapter draws on a body of qualitative analysis, comprising ten semi-structured interviews with activists and urban planning scholars. This work is part of fieldwork research that was conducted between 2014 and 2016 in Cairo and Alexandria. In addition, the analysis counts on auto-ethnography, given the author's personal involvement in the story. This ethnographic research is backed with an informed perspective on Egyptian law and the working of urban planning practice in Cairo.

Complexity and publicness in the Arab context

Public space is suffering from a great deal of reductionism under the umbrella of neoliberalism (Springer, 2010). The complexity thinking as an anti-reductionism approach, I argue, is therefore suitable to address contestation of public space. As such, I critically address two economic and political dimensions. In terms of the economic, neoliberalism submits the concept of public space to the dynamics of the market and economics. The ideology of neoliberalism assumes that social actors are driven by automatic mechanisms such as the 'hidden hand' and so on. Such assumptions reduce the multiplicity and value-loaded character of public space to the deterministic processes of the economy (Ravetz and Funtowicz, 1993). In terms of politics, it is intrinsic to neoliberalism to achieve a smooth flow of capital. To do so, governments that embark on this ideology strive for an ordered and harmonized public. Rancière (1999; see also Mouffe, 2006) contended that such attempts are reducing the dynamic space of 'political' difference to hegemonic order of the 'police'.

According to the second principle of classification, public space can be ranked as a complex system *par excellence*. Within the confines of its material space "the anarchy of the market meets the anarchy of politics" (Mitchell, 1995: 119). The success of a public space resides in its ability to intensify connections between diverse groups, inviting them to adapt and be aware of their differences and commonalities. Iconic public spaces, in particular, are value-loaded sites characterized by multiplicity and dynamism. They can either "stimulate symbolic identification, and cultural expression and integration" (Borja, 1998: 67) or manifest an ideology of exclusion, cultural clashes and segregation. Critical cultural geographers, and even political philosophers, tend to speak of public space in dialectic terms (Lefebvre, 1991; Springer, 2010; Rancière, 1999; Mouffe, 2006). Springer (2010) synthesized the two dialectic forces that influence public space in the context of neoliberalism. Those are the authoritarian force of the state to maintain order (or spaces of representation, in Lefebvre's terminology); and the informal force of the citizens to remain 'unscripted' (or representational spaces, according to Lefebvre) (Lefebvre, 1991). The continuous tension between these forces inclines us to look at public space as dynamic sites. In that respect, public space is always susceptible to disorder when one force 'temporarily' exceeds its antagonist. All these features of connectedness, diversity, adaptivity, multiplicity and dynamism are precisely what define a complex system.

Complexity thinking is a relational worldview. Addressing the Arab public space from this worldview helps in forming a contextualized understanding. Misconceptions within literature about the Arab context appeared due to a lack of such understanding. The first misconception is the habit of uncritically used imported concepts from the West to the social reality of the Arab world. The dichotomy between the Habermasian public sphere and the physical public space dominates much of the debate on the political role of public spaces and online media (Ben Moussa, 2013). For Habermas, deliberation and rational dialogue constitute the essential pillars of the public sphere concept. This dichotomy and the original concept of Habermas itself build on a culturally biased model of the difference between the public and private. Such a Euro-centric model stems from historical origins alien to the Arab context (Bayat, 2010). What counts as public or private in a Muslim-majority society emanates from a totally different herit-age. The religion of Islam actually challenges drawing sharp boundaries between the two realms (Ben Moussa, 2013).

In Cairo, I highlight two examples that sever the validity of imported models from the West. First, the *hara* – roughly translating as quarter – is one of the historical socio-spatial units of Cairo that persists till today. On the borders of the *hara* the state authority ends and all what lies between the buildings literally belongs to the dwellers of the area (of course, with inner power variations). Outsiders are called *aghrab* (foreigners), and cannot enter or use those spaces without the dwellers' consent – especially that of their respected elderly. Thus, those spaces are public to the dwellers but their publicness is conditioned by consent for outsiders.

Second, the division between the symbolic and the material aspects of public space is also derived from a Euro-centric perception. In parallel with the public–private dichotomy, there is not rigid distinction between the symbolic and the material when it comes to public space. In an interview with Khaled Al-Hagla in 2016 (an urban planning scholar who is based in Alexandria), he was asked about what obstructs a contextualized theorization of the Arab public space. He empha-sized that the weight of the symbolic in the Arab context is totally different from its European/Western counterpart. Regardless of which religion is embraced, the Arab citizen tends to glorify the symbolic aspect of a phenomenon much more than a European would do. In seven other interviews (2014–16) with diverse actors (activists and academics) in Cairo and Alexandria, the interviewees 'volun-tarily' chose to use metaphors, or make analogies with symbolic terms, in order to speak of features of public space, which supports Al-Hagla's claim.

Consequently, it is important to overcome such insufficient embeddedness of public space analysis, aiming to relate to the subjective notion of the public in the Arab context (Allegra *et al.*, 2013). Even before the events of the Arab Spring, urban scholars acknowledged this pitfall (Bayat, 2010; AlSayyad, 2004; Adham, 2009). This issue has led to the emergence of misleading *Orientalist* labels within literature about the public sphere of Muslim-majority countries. The most mainstream label is the 'Arab street'. This is a Western depiction of the public space and public sphere in the Arab world. This label implicitly entails treating the Arab context with a 'strong exceptionalist' attitude, picturing it as apathetic, immobile and unreceptive to modern democracy, and therefore categorizing it as a 'peculiar' context (Bayat, 2010: 3; see also Stadnicki *et al.*, 2014); its peculiarity lies in being culturally immune to freedom.

Another ramification of conventional Orientalism concerns the case of Egypt. The outdated divide between secularism and political Islam informs much of the analysis of the country's political landscape. This divide has been strengthened by local propaganda to invoke a sense of necessity for extensive security measures. However, the 2011 uprisings were spearheaded by different actors: those 'unsung heroes' the ordinary citizens (Shenker, 2016; Sims, 2010). This might explain why the recent Arab revolutions took all analysts and observers by surprise (Rabbat, 2011). The long-lasting belief in such misleading labels and outdated divides obviously hampers a nuanced understanding of contestation in the Arab cities (Allegra *et al.*, 2013). Conceptualizing this situation as path-dependent, the mainstream reductionist narrative was *self-reinforcing* for both local and global actors. At the same time, it limited the chances of seeing what lies beyond the path – and, hence, being totally 'surprised' when faced by unforeseen abruption. From this angle, the Arab Spring is seen as a catalyst for rethinking the analysis of urban contestation within the Arab-Islamic region (Stadnicki *et al.*, 2014).

Challenging the pitfalls of traditional (reductionist/deterministic) thinking, complexity science helps to derive a more contextualized and historically relevant understanding of contestation in urban Egypt. By tracing the path-dependent thresholds at which public spaces witnessed structural change, I aim to explore the structural shifts that caused the contestation – that is, to discover the roots of contestation.

Unfolding the roots of contestation

The structural condition of public space in a given moment is a guide to the relation between the state and citizens. Thus, to understand the structural changes of public space, it is important to know from whence the state derives its legitimacy. The Egyptian state mostly fits tentatively within the so-called 'soft states': one that presents itself as a fierce power that controls all aspects of development. But it notably lacks any institutional capacity to effectively execute or regulate those intentions (Dodge, 2012). Contradictory to its ostensibly neoliberal agenda (which entails the retreat of the state role in development), the Egyptian government acts, and perceives itself, as the central legitimate agent of development across the social, political and economic sectors. The market economy is exclusively captured among a network of power consisting of three major pillars: the crony economic and political elites; their trusted patronage circles of top-level bureaucrats; and the coercive iron hand of top police and military personals. Outside this 'power-triad' there is an extensive, apathetic and underpaid network of civil servants, who hardly grasp the tasks they are hired to do, due to the ambiguous overlapping between the different jurisdiction and executive layers (Sims, 2010).

What is mostly severed by this political landscape is the relation between the state and the ordinary citizen. He or she becomes accustomed to the fact that any fancy state-led development plan, no matter what rhetoric is used to convey it, is meant to serve the interests of *el akaber* – a word commonly used to refer to the elite oligarchs: those who have purchased their exclusive seats in the *polis*. Trust in official institutions is lost, and with it any belief in the commitment of the state to stand for the public interest. As long as this mistrust did not harm the stability of the established order of power, the state did not care much to restore the citizens' trust. Public space has been a projection of this relation on the ground. In public space, an observer can validate the consistency between the state actions and its claim to represent the public interest. An elected politician in parliament is supposed to represent the interests of his constituents. Likewise, public space represents the state's allegiance with the public interest (Mitchell, 1995). It is a sign of consistency or betrayal of the social contract that it claims to protect. In that sense, the interventions of planning with regard to space are, indeed, political. As an apparatus of the state, it is largely responsible for either delivering this consistency, or maliciously serving *el akaber*.

Despite of all these evident injustices and inconsistencies, the Egyptian state was considered by the international community as something close to a 'safety valve' in the region, due to its robustness and stability since it emerged in 1950s. Such an image, however, was remarkably disfigured by the domestic and regional transformations that took place from 2011. Many commentators have pointed out that public spaces of urban cities, in the material sense, played a paramount role in these socio-political transformations (Mitchell, 1995; Borja, 1998). However, prior to 2011 the impact of public space on politics was reduced to almost nothing. This means that the people of Egypt have been systematically deprived of their *right* to public spaces, as part of the state strategy of de-politicizing the public.

Over the past five decades, the state has relentlessly perfected those strategies, inventing new ones with one arbitrary goal in mind: to protect the status quo at all costs, which some see as the inherit 'violence from above' of neoliberalism (Springer, 2010; Borja, 1998). To do so, the state has devised official institutions and implicit strategies to perpetuate its control over the *symbolic* functions of public space, so as to legitimize any further constrains on the *material* public space. By exercising its power over both layers of public space, the state appropriated public space as a tool to forge the 'proper citizen': the obedient one who does not disturb their established order.

Following the relational approach embraced in this chapter, I start by dismantling the reciprocal relationship between the symbolic and material layers of the public space, paying attention to its *official* and *unscripted* statutes (Springer, 2010). Many Western, as well as Arab scholars have taken the material layer as their starting point to discuss the contested public spaces of the Arab Spring countries. Their fervour is understandable, given the euphoric and inspiring moments of the revolution in public squares such as Tahrir Square. Those moments are absolutely worth analysing on their merits.

However, as public space has more complexity to be accounted for, it is time to step back from such fascination with space and do more than just historicize particular events. With the aim of delving deeper in the subject matter, the symbolic layer is going to be my very point of departure to understand the forces that influence the material public space. Through an in-depth analysis of the Egyptian urban law and fieldwork research, I highlight four particular structural changes (institutional change initiated by the state) that shaped the boundary condition of the symbolic public space. The material manifestations in space are presented as illustrations of such structural changes.

Tracing path-dependent structural shifts

In this section, I analyse the structural change in Egyptian public spaces using the principle of path-dependency. As above, path-dependency as an approach to understand non-linear change is based on the idea that history matters. The analysis is made on a timeline that extends from 1958, in the post-colonial condition, till 2003, when internet connectivity was remarkably advanced. To operationalize the principle of path-dependency, the analysis is based on a model developed by the institutional theorist Bernhard Ebbinghaus (2009), summarized in four phases.

In the first phase, the path departs at a critical juncture at which the system is facing a societal crisis, or adapting to a radical change. Second, the path develops to a phase of institutionalization, going through self-reinforcing processes. Third, the path stabilizes itself to a sub-optimal path by "structuring the alternatives", meaning to limit the probability for different paths to emerge (ibid.: 195). Fourth, the path might go towards a departure phase. Once stabilized for a long time, the path may exhibit an accumulation of micro gradual changes, which could cascade macro systemic restructuring in response to contextual changes.

Next, I identify changes in the condition of the symbolic and material public spaces of Cairo, where the 'path' is theorized as the change in the structural condition of public space. At each threshold, I explain the socio-political condition that it creates and I identify its associated structural change in public space. In the logic of complexity, such thresholds are the bifurcation moments at which the system might evolve structurally to a different degree of complexity. The argument revolves around the idea that public space was incrementally stabilized by the state to a contested condition (sub-optimal path). In this condition, public space was deprived of two essential symbolic functions; (1) being supportive in identity formation; (2) making citizens feel they are recognized participants of the *polis*. In other words, this condition resulted in a crisis of identity and participation. How did this condition come to be? And what were the non-linear (unintended) consequences of this endeavour? These are the questions that I attempt to answer by tracing the *longue durée* of this crisis, theorizing it as a path-dependent process.

Figure 9.1 The heavy green and gold fences
Source: Mohamed El-shaheed, 2011

Confiscation of political life, 1958

The Egyptian socio-political structure experienced a crucial change in 1952. This year marked the independence from the monarchical and colonial rule. The military, under Nasser's rule, aimed to appropriate the country, launching a wholesale operation to nationalize any asset owned by the *ancien régime* and their associated aristocratic class. Based on the hierarchical ideology of the military, the newly emerging regime embraced a highly centralized system of governance (Kamel and Elhusseiny, 2014). Following the prevailing global trend in the 1950s, donors from the Eastern bloc and Western aid upheld state-led development plans in the post-colonial countries, aiming to earn their alliance. The Egyptian government made use of this support to actualize their dreams of modernizing Cairo. Modernization meant up-scaling the city's traditional fabric to car-friendly avenues and turning the public squares into nodes of traffic circulation. To ensure a smooth application of these urban geopolitical top-down plans, the central government solely captured all legislative and executive power.

Typically, such a political ideology does not tolerate any kind of competing opposition (Dabashi, 2012). Public opinion did not matter. Ordinary citizens were excluded from public decisions. In 1958, this exclusion was institutionalized. The state took a critical decision that would come to forge the political life and public space for the coming decades. The 'emergency law' was enacted, which led to the militarization of urban spaces and urban life in general (Abouelhossein, 2015). Cities and their large public squares were rendered as potential threat to the newly established order (Graham, 2010). In those spaces, citizens could congregate to express dissent, and this law gave the state legal authority to prohibit such assemblies. Besides, the new law gave the regime the right to control what gets published in all sorts of media. The symbolism of these practices is exemplified in the popular expression: *el hetan leha wedan* – the walls have ears (Abouelhossein, 2015).

This logic was imprinted in space by fencing off squares, subdividing them intro a smaller defendable enclaves. In a Foucauldian sense, such public spaces were turned into archipelagos, or "splintering urbanism" (Graham, 2010). The Haussmann-style street grid that was implemented during the monarchical era surprisingly facilitated such spatial strategies. This structural condition paralysed political life, and de-politicized public space. The symbolic functions of public space – to freely assemble in public – were constrained. By enacting the emergency law, the structure of public space was initially conditioned. Thus, it is plausible to theorize the emergency law as the *critical juncture* of the path.

Commoditization of public life in the 1970s

Following the triumph in the 1973 war in Sinai, Egypt was suffering from an economic crisis that exceeded the state's ability to cope. Since independence in 1952, the state was embracing a socialist mode in which the central government was responsible for all aspects of development. This was no longer possible. To get out of the recession, the Egyptian President Anwar al-Sadat advanced his alternative economic agenda of *infitah* (the opening). The word is almost self-explanatory; it means, literally, opening the economy to the free market and neoliberal imperatives. Of course, regional and international powers largely facilitated this transformation. In essence, the free market logic entails the liberalization of the economy. However, in the Egyptian context, this liberalization took a peculiar route. It was a 'conditioned liberalization' or 'liberal authoritarianism', in which the sovereign power could not be threatened (Dodge, 2012). The private sector funds that were meant to 'free' the economy were exclusive to economic elites. Only those who are included in the patron network of the ruling party were privileged. Public lands were captured by political and economic elites for high-end development projects (Stadnicki *et al.*, 2014). Cairo started to include the exclusive semi-public patterns of shopping malls and gated communities. Thus, this economic transformation of *infitah* gave rise to a new power grid and, in turn, transformed the city's spatial structure. The new structure was largely characterized by socio-economic inequality and class segregation.

Public space was already constrained by the emergency law. Upon embracing the neoliberal model, public space was further commoditized. The privatization of much of the city's public spaces sharply reduced any true public space. Citizens were denied spaces for identity formation and political expression (Kattago, 2012). Instead, public space was subjected to market mechanisms. At the core of the neoliberal spatial logic, an *ordered* version of public space is the most desirable. In this pro-control condition, capital can flow smoothly. It needs no disturbance from political expression and signs of dissensus. This logic was well aligned with the state's desire to consolidate power and monopolize the flow of capital.

The shrinking of the material public space reflects the shrinking role of ordinary citizens among the middle class and poor in public affairs. Also this reflects the shrinking of what Hannah Arendt calls the 'space of appearance' (Kattago, 2012): the space in which citizens' unique identities and values can be seen in public (Mitchell, 1995), some of which are at odds with the dominant politics. Without having a space to voice those alternative politics, they are destined to vanish. Consequently, by favouring ordered public space, neoliberalism is inherently anti-political (Springer, 2010). These conditions further strengthen the constrained public life established by the emergency law. Thus, neoliberal restructuring reinforced the initial path of the post-colonial condition. Incorporating the principle of path-dependency, the economic initiative of *infitah* could be theorized as a process of self-reinforcing the established path.

Stability as the dominant discourse, 1981

In 1981, the Egyptian economic and political landscape was taken over by the Mubarak regime. The dominant buzzword during this era was 'stability'. The entire regime – from Mubarak and throughout the ranks of his top bureaucrats – worked effectively to maintain the status quo. Both local and global powers counted on this regime to achieve this goal. Given its history and strategic geographic location, the stability of Egypt is fundamental to stabilize the geopolitics of the Middle East and North Africa region. In terms of economy, the regime built on the legacy of the *infitah* project. As the country was further integrated into neoliberalism, the geography of segregation and exclusion expanded the gap between the social classes. In terms of politics, the regime followed the model of 'democracy from above' or 'fake democracy' (Kamel and Elhusseiny, 2014). On the surface, it gave the image of representative democracy. But, in the corridors of power, election results were carefully directed to serve the interests of central government.

These economic and political mechanisms created a powerful network of elite oligarchs who believed they owned the country (Shenker, 2016). However, the regime realized, in order to reinforce their dreams of stability, they needed to extend the reach of their network. The regime embraced neo-patrimonialism as a mode of political rule (Dodge, 2012). In this mode, the regime forced citizens to base their survival on personal relationships with people in power. The relation between the state and society at large was made of hidden and informal networks. By doing this, the state mobilized a huge patronage network of actors to maintain stability.

Through controlling the patronage network, the state was able to define who got to be part of the symbolic public space. Becoming part of the patronage system meant being part of the exclusive *police* (Mitchell, 1995; Rancière, 1999). This condition spiralled down to the local level. Local politics were invested in an allegedly representative structure called the Local People Council (LPC). The LPC was supposed to convey the interests of local groups to the national policymakers. But, in reality, the process of appointing these councils was largely corrupt. Only local actors with close ties to the ruling party could become part of the council. Strong families in any local neighbourhood made sure to build those ties by a relation of mutual benefit. For example, such families offered great support to Mubarak's party during election time. These exclusive modes of politics killed the chance for meaningful political participation.

Asserting the dominance of the regime, the state envisaged a monolithic version of the public. The state used the formal media channels to manufacture a *norm* of the good citizen; one who is obedient and against the evil opposition, which tries to drag the country into chaos. Any form of difference was perceived as deviant and backward. Public spaces of the capital reflect this mode of governmentality. All iconic squares and major streets of Cairo's city centre were surrounded by heavy green and gold fences and subdivided by checkpoints, which made them fearful places for pedestrians. Such crude interventions were made by *al-Tansiq al-Hadari* (translates as systematic arrangement and arraying of urban space). This state apparatus emerged in 1993 and, ever since, public spaces have been under its management (Adham, 2009). As part of the authoritarian regime, it serves its interest in maintaining the status quo. The state used the symbolic and material public space to stabilize the exclusive policies towards public space. Hence, the condition of public space was stabilized into a sub-optimal path. Moreover, rejecting difference and alternative politics aligns with the third phase of Ebbinghaus's model of path-dependency. Rejecting difference is synonymous to the process of *structuring alternatives*.

Fusion with online public space, 2003

The year 2003 was one of contradictions. Geopolitical influences and internal structural shifts led to unintended consequences and non-linear change. On the one hand, the American invasion of Iraq in 2003 cascaded unforeseen results in Egypt. Following the chaos in the Iraqi socio-political arena, the Egyptian state strove to intensify its main source of legitimacy: stability. Mubarak's regime renewed its role as the natural custodian of order, making it clear in its rhetoric – claiming to have protected Egypt from similar chaos. This message was conveyed through tightening the security grip and deploying the emergency law to a higher degree (Dodge, 2012). This meant banning (slightly allowed) criticism of the government in televised and printed media. Checkpoints and security personnel became more numerous in public spaces. On the other hand, while this security campaign was underway, the Minister of Communication launched a contradictory campaign – focused on a hitherto insignificant medium.

Ahmed Nazif, the last prime minister before the revolution, was then the Minister of Communication. He was famously known for his campaign to develop internet connectivity and penetration in Egypt. During his term in office, Egypt Telecom made abundant deals with private companies to provide ADSL internet via landlines (Gerbaudo, 2012). As a result, internet penetration rose from 9 per cent to 24 per cent of households. Nazif's intended goal was to further modernize the state, opening the door for e-government and even smoother flow of capital. It soon irreversibly changed the structure of political society in Egypt. This change lies particularly in symbolic public space.

Before 2003, symbolic public space was successfully monopolized by the security apparatuses. The harsh lens of security had the capacity to filter what voices could be heard in public. However, the diffusion of internet, especially with rise of social media, severely challenged this capacity. Stories and images of oppression were circulated, discussed and criticized beyond the censored media. Once unwelcome voices were heard among the public, and resonated with like-minded citizens in remote areas. Through adopting policies such as the emergency law, the state was successful in scattering opposition voices. But ICT and social media opened new windows of opportunity for the oppressed voices, creating "new forms of proximity" (Gerbaudo, 2012: 13).

The relation between the material public space and online spaces did not appear as parallel or separate dimensions. In reality, interviewed activists described this relation rather as a fusion between the two. This fusion had its own structural features. According to Castells (2015), such interplay restructures collective actions. The produced structure takes the shape of the network. Public space that was once perceived as rigid structure and akin to a closed system co-evolved with online spaces; it dissolved into more a loose version, shifting towards an open system (Montuori, 2013). As explained in the third principle, this is a move towards a higher degree of complexity (De Roo, 2012).

Although the platforms of social media were censored by security, the 'normal' worldview could not relate to this increased complexity. Counting on the traditional methods of control, the state continued on its way, denying citizens the right to public space. The capacity of learning that typifies complex systems was largely ignored. The fusion rapidly empowered many new actors, networking enabling them to mobilize themselves and form coalitions. Within a year the first strong coalition appeared. In 2004, the movement Kifaya was established and utilized the internet in much of its collective actions. The movement presented itself as trans-ideological, which enabled it to unite almost all voices of opposition. Kifaya was the seedbed of 6 April Movement, which spearheaded the uprisings in 2011. Since 2003, online activism of movements such as Kifaya and 6 April has gradually melted the solid order of public space (Montuori, 2013). Throughout several waves of protesting, disorder repeatedly knocked on the door (Ben Moussa, 2013). The accumulation of those actions was restructuring the political society. That is to say, the self-reinforcing path was being challenged at its core: to protect order. Incorporating the model of path-dependency, this process is leading towards the fourth phase of *path-departure*. In a nutshell, the stabilized structure of public space was fundamentally shifting from 2003.

As seen from the previous thresholds, public spaces were subjected to the hegemonic force of ordering and arraying. Public space, by this definition, was a mere space of representation, projecting the state legitimacy (Lefebvre, 1991). The unscripted face of public space as representational space was diminished (Springer, 2010). Public space was solely given its 'official' face, which only accepts the obedient public. It excludes those 'different' identities that might be at odds with the dominant political ideology. To Arendt, the right to be freely different in public is a constitutive element of the 'human condition' (Arendt, cited in Kattago, 2012; see also Sennet, 1990). Thus, by denying them the right to be different, Egyptians were deprived of this essential condition.

Denying citizens the right to public space has created a crushed citizen. On the macro level, this condition led to the loss of civic pride. The majority of citizens were stigmatized and suffered from a long history of being *capitis diminutio* (Borja, 1998). The forces of securitization and neoliberalism created distorted public spaces. Ideally, public space should be a dynamic site of active citizenship (Springer, 2010). Forces of ordering flattened this dynamism. Citizens were only permitted to use public space passively (Bayat, 2010). Public space did not function as a place for citizen participation, in which citizens could voice their dissent. Participative uses of space were considered assaults on state authority. Only the properly behaved citizen was allowed to appear in public: the one who conforms to the "ways that the state dictates" (ibid.: 11).

However, the fusion between the material and online spaces generated new affordances. Within this new structure, citizens managed to compensate for the missing functions of public space. This structural shift empowered oppressed voices. The silenced resistance became tangible contestations through movements such as Kifaya and 6 April. With this turn, the landscape of collective actions became too complex to contain using the traditional ways of the state – ways that were designed for the old version of a closed/static public space. But the formidable condition of the material public space kept on reminding citizens of their powerless status and lost civic pride. It was not until the social upheavals of 2011 that civic pride was temporarily restored. It was restored when ordinary citizens resorted to occupying the public squares, forcing the state to consider their active participation and to recognize their different identities.

Conclusion

Up until the Arab Spring events, public space – as an issue for contestation – was absent from much of the socio-political theorization of the region, mainly due to reductionist and Orientalist labels. Such labels do not do justice to the contextual and historical layers of Arab socio-political reality. The long belief in these labels was violently shaken by the Arab revolutions of 2011. Nobody saw this coming, let alone knew where it came from. Tackling these theoretical pitfalls, this chapter pictured these events as an opportunity to 'finally' go beyond the stereotype and bring about new lines of theorization in light of contextual factors. With this aim, anti-reductionist and relational approaches of complexity were incorporated. Using these approaches, the contested public spaces of urban Egypt were anchored in a socio-historical and urban geopolitical analysis. I argued that the roots of contestation could be deduced through conceptualizing how it emerged as a path-dependent process. By delving into the roots of contestation, this chapter illustrates that contestation emerged due to the changing structural condition of public space. From a complexity perspective, this change develops alongside path-dependent thresholds – that is, by deploying the principle, 'time' was brought to the centre of the analysis. The roots of contestation were traced on a timeline that extends from 1958–2011. Along this trajectory, I identified four thresholds at which the structural condition of public space notably changed. This change was associated with political, economic and techno-social structural shifts in the Egyptian society. First, the post-colonial regime enacted the emergency law in 1958 that begun to constrain the symbolic functions of public space. Public space lost its status as a place for the active participation of ordinary citizens.

Second, this constrained condition was reinforced by integrating the country in the economic agenda of neoliberalism, following the state desire to maintain an 'ordered' public space so that capital could flow smoothly. As such, public space was surrendered to corrupt market dynamics, and citizens were denied the right to a free public space. Third, since 1981 Mubarak's regime embraced 'stability' as their dominant discourse, manifested by the monolithic and static condition of public space: one that does not tolerate dissent; one that accepts only passive and obedient usages of space. The analysis concluded that this stabilized condition resulted in a deep-rooted crisis of participation and identity.

Fourth, the advancement of the internet since 2003 fundamentally shifted this rigid structure. This shift afforded the oppressed political identities the opportunity to voice their demands and build coalitions. The new affordances of online space compensated for the missing functions of public space. I suggested that the fusion between online and material spaces increased the complexity of public space. This condition turned contestation from an invisible to tangible situation: a process that kept on building up till it contributed to the social upheavals of 2011. Due to the spread of social media, public space became more open, more complex and susceptible to disorder. Thus, the online and physical public spaces co-evolved into something new. This newness invites new lines of theorization; and this chapter is a modest attempt to ignite this line.

References

Abouelhossein, S. (2015) Urbanism of exception: reflections on Cairo's long lasting state of emergency and its spatial production. Urbino (Italy) Research Committee 21, 27–29 August.

Adham, K. (2009) Making or shaking the state. In D. Singerman (ed.), *Cairo Contested: Governance, Urban Space, and Global Modernity*. Cairo: American University in Cairo Press: 41–61.

Allegra, M., Bono, I., and Rokem, J. (2013) Rethinking cities in contentious times: the mobilisation of urban dissent in the 'Arab Spring'. *Urban Studies*, 50(9): 1–14.

AlSayyad, N. (2004) Urban informality as a 'new' way of life. In A. Roy and N. AlSayyad (eds), *Urban Informality: Transnational Perspectives from the Middle East, Latin America and South Asia*. Lanham: Lexington: 7–30.

Bayat, A. (2010) *Life as Politics: How Ordinary People Change the Middle East*. Amsterdam: Amsterdam University Press.

Ben Moussa, M. (2013) From Arab street to social movements: re-theorizing collective action and the role of social media in the Arab Spring. *Westminster Papers in Communication and Culture*, 9(2): 47–68.

Borja, J. (1998) *Citizenship and Public Space* [online]. Available from: www.publicspace. org/en/text-library/eng/11-ciudadania-y-espacio-publico. Accessed 20 August 2016.

Castells, M. (2015) *Networks of Outrage and Hope: Social Movements in the Internet Age*, 2nd edn. Cambridge: Polity Press.

Dabashi, H. (2012) *The Arab Spring: The End of Postcolonialism*. London: Zed.

De Roo, G. (2012) Spatial planning, complexity and a world 'out of equilibrium': outline of a non-linear approach to planning. In J. V. W. Gert De Roo (ed.), *Complexity and Planning: Systems, Assemblages and Simulations*. Farnham: Ashgate: 141–75.

Dodge, T. (2012) *From the 'Arab Awakening' to the Arab Spring: The Post-colonial State in the Middle East*. London: IDEAS reports, London School of Economics and Political Science.

Ebbinghaus, B. (2009) Can path dependence explain institutional change: two approaches applied to welfare state reform. In J. O. Lars Magnusson (ed.), *The Evolution of Path Dependence*. Cheltenham: Edward Elgar: 191–212.

Gerbaudo, P. (2012) *Tweets and the Streets: Social Media and Contemporary Activism*. New York: Pluto.

Graham, S. (2010) *Cities Under Siege: The New Military Urbanism*. London: Verso.

Kamel, B., and Elhusseiny, M. (2014) Dynamics of social sustenance in Cairo's changing public space. *Journal of Applied Social Science*, 6(10): 471–95.

Kattago, S. (2012) How common is our common world? Hannah Arendt and the rise of the social. *Problemos*, 81: 98–108.

Lefebvre, H. (1991) *The Production of Space*. Oxford: Blackwell.

Mitchell, D. (1995) The end of public space? People's park, definitions of the public, and democracy. *Annals of the Association of American Geographers*, 85(1): 108–33.

Montuori, A. (2013) Creativity and the Arab Spring. *East–West Affairs*, January–March: 30–47.

Mouffe, C. (2006) *The Return of the Political*. New York: Verso.

Portugali, J. (2006) Complexity theory as a link between space and place. *Environment and Planning A*, 38: 647–64.

Rabbat, N. (2011) Circling the Square. *ARTFORUM*, April: 182–91.

Rancière, J. (1999) *Disagreement: Politics and Philosophy*. Minneapolis: University of Minnesota Press.

Ravetz, S., and Funtowicz, J. (1993) Science for the post-normal age. *Futures*, 25(7): 739–55.

Sennet, R. (1990) *The Conscious of the Eye: The Design of Social Life and Cities*. London: W. W. Norton & Co.

Shenker, J. (2016) *The Egyptians: A Radical Story*, International Edition. London: Penguin.

Sims, D. (2010) *Understanding Cairo: The Logic of a City Out of Control*. Cairo: American University in Cairo Press.

Springer, S. (2010) Public space as emancipation: meditations on anarchism, radical democracy, neoliberalism and violence. *Antipode*, 43(2): 525–62.

Stadnicki, R., Vignal, L., and Barthel, P. (2014) Assessing urban development after the 'Arab Spring': illusions and evidence of change. *Alexandrine Press*, 40(1): 1–9.

Tierney, T. (2013) *The Public Space of Social Media: Connected Cultures of the Network Society*. New York: Routledge.

Part IV

Urban geopolitics

Latin America

Camillo Boano

The fourth part focuses on Latin America and attempts to offer a glimpse into the complex and profound urban geopolitical challenges faced by urban areas. This is far from complete or exhaustive. It aims to reflect on three Latin American cities, each with distinct urban processes showcasing the need to diversify the understanding of local urban geopolitics. The urban debates around Latin American urbanism are known to be connected to such phenomena as gentrification, metropolization and informalization, as well as sharp social and spatial inequality. The three chapters aim to connect some of those key urban debates in a geopolitical dimension. Camila Cociña and Ernesto López-Morales explore gentrification processes in relation to the emergence of non-violent conflict in Santiago de Chile. The authors argue that exploring the notion of conflict as an essential part of democracy is an opportunity to shift urban geopolitical power struggles through local mobilization of community groups. Catalina Ortiz and Camillo Boano discuss Medellín's aspiration to consolidate as global benchmark of urban innovation, reflecting on the production and rearrangement of urban space driven by the urban geopolitics of informality, the state-led and municipally governed super accelerated and highly aesthetical infrastructure development and the politics of securization and control. In the closing chapter, Peter D. A. Wood calls for a more critical attitude and geopolitical approach to urban studies, discussing a methodologically distinctive perspective measuring trans-border participation through urban development planning in Foz do Iguaçu, Paraná, Brazil.

10 Unpacking narratives of social conflict and inclusion

Anti-gentrification neighbourhood organization in Santiago, Chile

Camila Cociña and Ernesto López-Morales

Introduction

Historically, Latin American cities have developed patterns of growth strongly shaped by socio-economic segregation (Borsdorf, 2003). Santiago is not an exception to this rule as it is currently considered one of the most segregated cities of the OECD members (OECD, 2012). Even though there have been important efforts from the Chilean state in ensuring access to land and housing to low- and middle-income households (policies that are considered successful in quantitative terms; see Gilbert, 2002), over the last 30 years the logic of property-led exploitation of land and the private extraction and accumulation of its value have pushed poorer social groups outside the city, where low quality residential environments are built (Ducci, 2000; Rodríguez and Sugranyes, 2005; Sabatini *et al.*, 2001; Sabatini and Brain, 2008). More recently, this pattern of growth has somewhat changed, as middle-income housing markets now have a greater preference for inner city areas, both perpetuating and transforming the traditional logic of segregation and division in Santiago, as this arrival also generates gentrification of low-income inner quarters (López-Morales, 2013).

In Santiago, over the last few decades, there has been a decreasing number of households living in derelict parts of the inner city. However, according to the 2002 National Census (INE, 2002), by that year and despite the decrease, the number of people still living in the 11 inner districts of Santiago was approximately a quarter of the city's population. These households are still mainly composed by traditional low-income citizens and, more recently, by international migrants (Margarit and Bijit, 2014). Now they face threats of displacement through gentrification processes for the rapid and sometimes intensive redevelopment of their hitherto affordable neighbourhoods, and the massive influx of middle-income households with higher consumption power and cultural capital (López-Morales, 2011, 2016; López-Morales *et al.*, 2015: Contreras, 2011; Inzulza-Contardo, 2012).

This chapter aims to briefly describe the emergence of new spaces of coexistence and conflict in Santiago, amid current trends of intensive redevelopment. We show here how class-related power relations emerge and are reproduced, often ending in the radical transformation of neighbourhoods and, ultimately, generating displacement as a frequent outcome of this spatial conflict. However, as new evidence arises, this chapter asserts that conflict resolution and coexistence is also possible, as some traditional local communities struggle to avoid their displacement from their inner city locations. We contend that the emergence of spaces of class-related conflict can be understood as opportunities for the encounter of clearly differentiated positions, as groups that usually don't see each other in segregated cities as Santiago can find the space for expanding democratic exchanges, via collective organization and struggle deployed by the less powerful actors of the city. In fact, this can be a way to increase the room for manoeuvre of the less powerful groups, if the emergence of conflict and conflict resolution is understood as a tool for opening differences and deepening democracy (Mouffe, 2005; Miessen, 2010; Swyngedouw, 2011).

To substantiate this argument, we focus here on three cases: first, the achievements of the *Inmuebles Recuperados por Autogestión* (IRA), which, from 2012–14, managed to sustain provisional low-cost housing in spaces undergoing gentrification, through the implementation of a social real estate agency; this has happened after years of highly organized grassroots activism. Second, we examine the case of the *Consejo de Movimientos Sociales de Peñalolén* (CMSP), which was able to organize opposition to a local-level land-use master plan, shifting the development path of the whole district. And, third, we review the case of *Italia-Caupolicán*, a neighbourhood where commercial and programmatic transformations have brought new social groups to the area, and in which the importance of traditional identities and heritage have allowed low-income actors to remain and play a key role in the transformation of their neighbourhoods. The cases are discussed in their capacity of producing urban geopolitical changes from grassroots organizations, using the emergence of conflict and conflict resolution as a way to challenge segregating and unjust patterns of urban development. All cases have unfolded different spatialities in the city, which are also described here. Despite its contextual differences, these cases can also be seen as urban geopolitical changes 'from below'.

We use the idea that conflict emergence and resolution is an essential component of truly democratic processes (Mouffe, 2005; Badiou, 2013), and explore the implication of this notion in urban development, particularly in allowing the emergence and encounter of different conflictive voices in contested urban transformations (Miessen, 2010; Swyngedouw, 2011; Cooke and Kothari, 2001). We aim to reflect on the underlying forces behind processes of consolidation of new ways of exclusion and inclusion in these areas. We provide several reflections at the end of the chapter to hint at how gentrification, understood as a clear-cut conflict, can be tackled by well-organized, conversational communities, even despite the existing class differences among residents. This is done in order to tackle, avoid or reject gentrification from the neighbourhoods, and promote inclusion. Our approach also contests the erroneous idea that gentrification, namely the influx of upper-income users into a neighbourhood, means exclusively a process of displacement and exclusion *beyond* the control of existing local agents and communities. As we see next, gentrification can be tackled by social autonomous organization in some neighbourhoods, and also can be channelled by increasing social participation in urban decision processes. We do not believe gentrification is good because it increases political power among gentrified communities *per se*, but we see here that this avoidable problem can trigger new forms of participation and self-management of the neighbourhoods. Therefore, as we will explore in the following sections, gentrification processes can imply local geopolitical transformations that shift the development of certain neighbourhoods, and translate into contested city projects; power conflicts are central in the resolution of those possible development paths.

Gentrification as the industry of exclusion and the role of social conflict as a means for inclusion

During an informal conversation, the sociologist and urban planner Jorge Fiori once drew attention to the song 'Años' by the Cuban songwriter Pablo Milanés, to refer to the idea of *consensus* and the importance and value of *conflict* in political and urban processes.

The song, which talks about the pains of a couple getting old, says, "You say yes to everything, I say no to anything, to forge this great harmony, which makes hearts old" (Milanés, 1980). The *great harmony*, as opposed to the spaces of conflict, would be the source of struggle for the old couple, as it can be for contemporary democracies and societies. What great harmonies usually hide is that some voices are unable to challenge processes led by stronger ones, whose power lies either in their capital control or political hegemony.

Many authors have identified the notion of *conflict* as a productive approach to deepen democratic processes (Badiou, 2013; Mouffe, 2005; Miessen, 2010; Swyngedouw, 2011). Furthermore, non-violent conflict, as opposed to consensus, is presented as a necessary element of any true democracy. In practice, however, conflict needs to take place in contexts in which different voices can actually interact on a relatively equal footing. We know that most urban processes occur in the context of uneven power relationships, where not everyone can play an equal role in what Sandercock calls the "urban conversation" (2003: 220).

The idea that cities should be the space for inclusion and sites for difference has been at the core of most current literature looking for urban justice (see Madanipour, 2007; Fainstein, 1999, 2010; Sandercock, 2003). But in order to build more integrated neighbourhoods it is necessary not just to bring together different groups, but also to understand what it means to live with difference, giving space for conflict to emerge, for different voices to be heard and to make those differences flourish – until not, perhaps, a consensus, but at least where certain conditions or goods that benefit the less powerful or affluent ones are achieved.

From a political theory perspective, Chantal Mouffe (2005) has discussed the notion of the impossibility of a world without antagonism, and therefore the necessity of the existence of clearly differentiated positions and political alternatives as constitutive elements of modern democracies. In his urban interpretation of Mouffe's work, Markus Miessen uses her framework, stating "any form of participation is already a form of conflict" (2010: 91). Erik Swyngedouw (2011), in his book *Designing the Post-Political City and the Insurgent Polis*, introduces the idea of the 'post-political consensus' city, where neither democracy nor politics are possible, as neoliberal consensus drowns any possibility of the emergence of conflictive spaces, bringing about difference and diversities.

The idea of politics as something that is only possible through conflict, and Swyngedouw's acknowledgement of the lack of those conflictive and contested spaces in current politics, suppose very particular challenges in understanding the role of gentrification processes, as spaces in which non-violent conflict can emerge very clearly, and where, if allowed, differences can be part of the urban conversation, and become very productive in increasing integration in cities.

Gentrification's main conflict is the threat of displacement; the latter means when wealthier people displace poorer people, where social and cultural diversity is replaced by homogeneity, and this undermines the chances of a neighbourhood as conversational place (Lees *et al.*, 2016). Displacement means considerably more than physical expulsion because it encompasses phenomenological displacement or exclusion from decisive political spheres (Davidson and Lees, 2010). In López-Morales (2016) and López-Morales *et al.* (2015), gentrification is defined as a set of outcomes of an array of local and global factors, which lead to the loss of use value and cultural and social capital set in space, as well as the denegation of the right to use certain space and participate in the spheres of social reproduction, for the many at the bottom of society, as soon as the neighbourhood space is revalued by wealthier and more powerful users. The latter impose their specific class-related demands on space, to the point that any other value consideration or expectation from different actors becomes irrelevant for those who make unilateral decisions about the redevelopment of the neighbourhood.

The previous reading is largely inspired by Eric Clark, who defined gentrification from a generic point of view as "a change in the population of land-users such that the new users are of a higher socio-economic status than the previous users, together with an associated change in the built environment through a reinvestment in fixed capital" (2005: 258). In Chile, gentrification has been led by those 'users' with enough power to either transform the space where they live, or where other people live, while 'a change in the population' is displacement, both physical and phenomenological (Davidson and Lees, 2010; López-Morales, 2016; López-Morales *et al.*, 2015; Janoschka *et al.*, 2014).

However, some authors argue that the phenomenon responds to a middle-class agency on housing and urban redevelopment that reduces segregation by socially mixing the neighbourhoods, not necessarily producing displacement, and improving the physical fabric of the place (Vigdor, 2002; Freeman, 2006). In the case of Chile, seen mainly from the perspective of the newcomers' preferences regarding lower-class existing residents, gentrification has been seen as a socially inclusive process (Sabatini *et al.*, 2009), even though it responds to a whole process of urban reconfiguration going on in the city, led by a powerful private real estate sector (Janoschka *et al.*, 2014).

Santiago de Chile: real estate and commercial gentrification as urban phenomenon

Since the late 1970s, Chile has implemented housing policies based on subsidy provision, in which the private sector is in charge of the production, management, construction and capture of profit of housing production. The scenario after three decades of such a private-led model has different faces. On the one hand, a successful finance system structured by subsidies reduced dramatically the housing deficit, providing an average of 90,000 subsidies yearly between 1990 and 2000 (Salcedo, 2010) – meaning that, by 1993, "a Chilean-type model, or at least elements of the Chilean model, had become acknowledged best practice" (Gilbert, 2002: 310). On the other hand, the positive numbers contrast with massive displacement produced by deregulated production of segregated and sprawling cities, driven mainly by the will of reducing costs and increasing profit by private companies, through the production of low quality standardized solutions, on cheap land and without basic equipment and connectivity, reinforcing and consolidating segregation patterns (Ducci, 2000; Rodríguez and Sugranyes, 2004, 2005; Sabatini *et al.*, 2001; Sabatini and Brain, 2008; Salcedo, 2010).

Adding to the construction of social housing on cheap land, Santiago's territorial coverage has grown due to a series of phenomena: high-income groups, historically concentrated in the north east of the city, have continued colonizing the mountains in that direction; policy instruments and infrastructure investment have allowed the construction of private suburban gated communities on the outskirts of Santiago, in areas such as Talagante, and particularly in the province of Chacabuco (Vicuña del Río, 2013). Emerging middle classes have also urbanized areas in low-density schemes around sub-centres such as Maipú and La Florida. Beginning in the early 1990s, however, Santiago has seen a back-to-the-city movement, with increasing high-rise residential redevelopment, attracted by tax incentives, aimed at middle-income strata in certain central neighbourhoods with high transport connectivity. A second back-to-the-city trend has been the more specific: slower, but with conspicuous colonization of traditional neighbourhoods by middle classes, where well-educated new residents bring in new heritage protection agendas.

In all cases, those areas have shown large rent gaps that now come to be captured by more affluent users, basically private developers (López-Morales *et al.*, 2012). In the case of Santiago-Centro *comuna*, this process has also shown a deep demographic change and a complete reversal of the traditional centrifugal pattern of development experienced by Santiago during the second half of the twentieth century. The 2012 National Population Census showed 43.8 per cent growth between 2002 and 2012, while previously, between 1992 and 2002, this *comuna* lost 9.4 per cent population (INE, 2002, 2012);[1] namely, while, by years, the city centre lost population, over the last decade a flow of new residents arrived in the inner city.

However, this repopulation of the city centre has taken place in an area that was never empty: although it has been previously documented that many of these central areas were depopulated (Arriagada *et al.*, 2007), according to the 2002 Census these *comunas* still hosted around 1.3 million people, roughly 25 per cent of the city's population. The socio-economic composition of the remaining population of the city centre was basically low- and lower-middle income strata (INE, 2002).

Rojas (2004), Arriagada *et al.* (2007) and Contrucci (2011) were influential policy advisors during the 'golden period' of high-rise redevelopment in the 1990s as they saw the high-rise residential market in the central *comunas* as an opportunity for demographic repopulation and social mixing. They claimed that higher income consumption power held by the newcomers would trickle-down and so help 'regenerate' local environments. In fact, they were responding to very powerful discourses of the time that tended to blight central areas as derelict, denying the fact that, among depopulation and physical deterioration, many renewing urban areas had hosted vibrant local communities. At the time, little attention was paid to the actual effects as perceived and experienced by the original lower-income population who resided in those neighbourhoods. Serious accounts of displacement are much more recent and critical about the loss of opportunities for low-income residents that stay or those forced to move in redeveloping central neighbourhoods (López-Morales, 2011, 2013; López-Morales *et al.*, 2015; Contreras, 2011). In addition to the high-rise residential renewal triggered by tax incentives and subsidies, high-income groups, particularly young professionals, started occupying some heritage central neighbourhoods such as Lastarria, Brasil, Italia-Caupolicán and Bellavista (Contreras, 2011; Schlack and Turnbull, 2009, 2011, 2015).

This process of central densification has implied flows of population through gentrification that have encountered low-income people who remained living in derelict central quarters. In this conjuncture, some conflicts have emerged, probably strengthened by the fact that, historically, the city of Santiago has rarely allowed the encounter of different social groups. The cases examined hereinafter are examples of how gentrification processes and conflict resolution can challenge these dynamics and traditional top-down housing policies, in the power struggle for determining the future of local urban geopolitical conditions.

Narratives of social conflict and inclusion in three neighbourhoods of Santiago

In this section we will explore three cases that wrestle with the theory presented above: the first two discuss the role of the current *pobladores* movements in Santiago-Centro and Peñalolén *comunas* that have effectively managed ways to prevent social displacement from areas upon the arrival of social groups and real estate investors and redevelopers with higher consumption power into their spaces; the third case describes the different spaces of encounter that emerged in Italia-Caupolicán, a heritage neighbourhood where changes from residential to commercial use have brought new social groups even without the strong presence of real estate development. In each of the cases, we explore the new spaces of coexistence, trying to understand the room for conflict resolution in which different voices can

Figure 10.1 Landscape of high-rise renewal in Santiago
Source: Daniel Meza, 2011

take place in a democratic urban conversation and production, ensuring low-income actors play a key role in their neighbourhoods' transformation processes.

Anti-gentrification grassroots organization to tackle gentrification by real estate investment: the case of IRA

As currently part of the *Movimiento de Pobladores en Lucha* (Movement of Residents in Struggle; see Renna, 2014), the IRA network, since 2010, has spread through some of Santiago's inner neighbourhoods: a clear-cut demand by tenants living in Santiago-Centro *comuna*. Originally, these households were mainly low-income informal or blue-collar workers or street traders, all fighting for the right to stay in the neighbourhoods after the dwellings they occupied were severely damaged – an effect of the 2010 earthquake, which shook a large number of old dwellings in central Chile, many of which were declared derelict by municipal officials, thus dangerous to live in.

Figure 10.2 Location of cases in Santiago
Source: Authors, 2016

The 2010 earthquake indirectly led to a speculative rise in renting prices, and this created displacement pressure among homeless tenants. In 2011, this *pobladores* assembly established a strategy for property occupation in the Santiago-Centro *comuna*. Two private and one public property were squatted in; those had been abandoned for decades by their owners and were located amid very central, renewing neighbourhoods. IRA saw these sites as 'transitional housing' for the several tens of households occupying them; they would be also a political means of generating permanent public and media interest insofar as proper housing solutions were not given by the authority.

All in all, the claim from the IRA was related not only to critically unveil a situation of emergency and catastrophe suffered by hundreds of tenants, but also to actually produce (temporary or not) a popular habitat in the very central quarters of Santiago, a 'locational' demand that, for the last 40 years, the Chilean housing policy had not met. IRA's claim was to retain and regain centrality from a class-differentiated agent perspective; this is directly opposed to the commodity production by real estate private markets that characterizes the current urban renewal policy. The Chilean government responded, first, by stigmatizing these squats, labelling them as against the defence of the right to private property (of land), which is granted in the 1980 Chilean Constitution; regarding the public properties occupied, the government reinitiated a process of bidding to reinstate private proprietors (but not *pobladores*) over these sites.

In the meantime, IRA experienced remarkable solidarity from other grassroots organizations, neighbour residents and *pobladores* from other areas throughout the city, which were facing threat of displacement. At present, the assembly of the IRA works with support from the Ministry of Housing and Urbanism on a public housing project for 50 households, which will smash and renew one of the properties previously squatted, namely the Casona Esperanza. The state support for IRA's project now consists simply of the issuing of housing subsidies to the settlers. Yet, the IRA assembly has not had an easy time, as the building process has been delayed by substantial bureaucracy from/between incumbent state bodies, increasing, as well, the overall project budget.

This case shows how residents facing imminent displacement – triggered, in this case, by the earthquake and the subsequent processes of speculation – find strategies to increase their capacity of negotiation, creating spaces of resistance that allow them to gain leverage for remaining and renewing their original neighbourhood, and producing a habitat for low-income groups in a very central area of Santiago, thereby tackling the growing problem of homelessness and displacement by the earthquake that existed in the *comuna*.

Social movement articulation in Peñalolén: the social and political struggle for the management of land for social housing

The second case is the *Consejo de Movimientos Sociales de Peñalolén* (Peñalolén Council of Social Movements, CMSP). The CMSP was, in fact, a coordination of several organizations born in the context of the grassroots opposition to the approval of a new land use regulatory plan for Peñalolén that was being implemented by the municipal government from 2009. If it were approved, the proposed plan would have up-zoned several areas with increased floor area ratios, thus permitting higher-storey residential densification similar to what is shown in Figure 10.1.

Peñalolén *comuna* is located in the eastern peri-centre (between the core and the periphery) of the city, with a grassroots character as its territory consisted of poor squatted land and temporary settlements (later on formalized) in the mid-1950s. However, from the 1990s onwards, this *comuna* started to see intensive construction of new residential gated condominiums by private real estate firms that were meeting upper-income housing demand. In this changing context,

Figure 10.3 Leaflet calling for support to one IRA's house
Source: David Kornbluth, 2014

Mayor Claudio Orrego started to redraft the *comuna* regulatory plan. The redevelopment goals of this new plan would impose stricter building conditions for the low-income housing in Peñalolén; it would also widen the main roads with a series of forced purchases of inhabited land by the state, and it would give more attractive building conditions to the property-led, high-rise residential market.

In order to tackle this situation, Peñalolén's organized tenants and other social and ecological-protection organizations formed the CMSP and then brought into question three issues related with the new plan: namely (1) the inconvenience of up-zoning the land to attract high-rise construction, (2) the imminent imposed purchase that numbers of landowners and tenants would face for being located aside the main roads and (3) the proposed new community park, which was located in a place where a slum already existed, thus forcing the displacement of hundreds of settlers. The claim by the *pobladores* of Peñalolén was not only focused on a defence of their quality of life (rights to sunlight, privacy and certain controlled density), heritage or environmental conservation, but also they demanded the use of peri-central land for low-income housing construction, rather than using it for the expansion of an upper income-oriented housing market.

The CMSP managed to gather 5,300 signatures to convene a community referendum. This was the second time in Chile's post-dictatorship democratic history that a local-level referendum was called by grassroots organizations. After a hectic competitive campaign by the CMSP and the Peñalolén municipality, on 11 December 2011 the referendum took place and the option of 'reject the Communal Regulatory Plan proposed by the Mayor' won, with 52 per cent of the votes. From then on, the CMSP managed to contain the implementation of the Regulatory Plan and thereby increased its level of both social and political power in the *comuna*.

This case shows that social and political struggle was fundamental in order to improve some conditions for the land management of social housing, an issue that was almost impossible before, given the increasing land prices in this *comuna*. Second, the case shows the achievement of a rearticulation of the movement of *pobladores*; from then on, the CMSP presented some of its leaders as candidates to the municipal council, gaining one representative position for the period 2012–16.

Commercial gentrification in heritage areas: building common identities in Barrio Italia-Caupolicán

Italia-Caupolicán[2] is a residential neighbourhood in the wealthy central *comuna* of Providencia; centrally located, it is composed mainly of one- and two-storey houses and some industrial infrastructure built at the beginning of the twentieth century, embracing a rich architectural heritage. Historically, it has been one of the poorest areas of the *comuna*, with some degraded infrastructure and a concentration of elderly persons, migrants, furniture-makers, small-scale mechanics and craftsmen. Over the last ten years, however, it has witnessed a transformation – mainly led by young designers and artists moving their business to the area. Linked to the historical presence of furniture-makers and antique dealers, new design stores, restaurants and artists' workshops moved to the area – first, as independent initiatives and then, over time, in middle-scale redevelopment transformations of old houses into commercial galleries.

The gentrification process here has been led mainly by a change of use, from residential to commercial; according to Schlack and Turnbull (2015), the population decrease in the neighbourhood over the last few years can be explained partially by this change: 92 per cent of the 48 properties that were brought into commercial use during the period 2009–12 were previously residential buildings. This change of use has not been accompanied by an aggressive change of the urban infrastructure; the profitable potential of properties has been enlarged by the change of use rather than by the multiplication of it, as the "value is being found in the rehabilitation of existing structures and not in new-build, high-rise redevelopment" (ibid.: 354). In that sense, Italia-Caupolicán could be facing what Holm calls *displacement by distinction* (in Schlack and Turnbull, 2015), where new identities colonize neighbourhoods through different forms of urban investment, dominating the construction of collective narratives and transforming territories to the point of becoming culturally hostile spaces for former residents.

Distinct from the two cases presented above, the mechanisms used by traditional neighbours of Italia-Caupolicán haven't been centralized and have rather taken a variety of forms in facing the threat of displacement, becoming key actors in different instances of the construction of old and new identities within the area. The change of use from residential to commercial, and the consequent change in the profile of neighbourhood visitors, has implied conflicts in terms of the contested construction of a common identity. The question is to what extent the voices of former residents have been part of the creation of this new identity; or, in other words, which instances have allowed the emergence of spaces of conflict in which less dominant voices are able to participate?

The most evident friction arises from the differences between the traditional occupants of the more residential southern area of the neighbourhood and those who inhabit the more commercial sectors, both as residents and as commercial entrepreneurs: residents who have been in the area for a longer time feel that they do not 'belong' in the neighbourhood, and actually they call themselves as part of 'Barrio Santa Isabel' rather than Italia (Melillán and Cruz, 2014).

This active differentiation from some of the former residents, however, is not reciprocal, as the physical and cultural heritage existing in Italia-Caupolicán is a key asset in the creation of value for new entrepreneurs, as is evident in the language and images presented by *Corporación Barrio Italia*, constituted in 2008 by a group of entrepreneurs of the area. This corporation states that they are an initiative "founded and driven initially by traders… that then added the residents and different actors of the neighbourhood to a common project" (Corporación Barrio Italia, n.d.). This change, from an initially sectorial initiative to one that includes residents as actors, shows the necessity of inclusion of inhabitants of Italia-Caupolicán as part of identity-construction.

In 2012, a new mayor was elected in Providencia *comuna*; one of the main focuses of her campaign and new administration was residents' participation in communal processes, in stark contraposition to the authoritarian previous mayor. Official participation processes took place in the area, enabling a progression towards changing the communal planning, protecting the physical heritage of the area. This regulatory framework has been possible, in part, due to the organization of residents, showing the conflictive nature of the changes that were taking place in the area.

However, the transformation of old houses into commercial galleries has taken place mainly through traditional expulsion processes in which negotiation occurs between the property owner and the potential investor and developer. Even if the built heritage is typically respected and used as part of the value, the spaces for conflict and negotiation are very much constrained and controlled by capital interests: the population in the supposedly most gentrified area of Italia-Caupolicán decreased by 56 per cent between 1982 and 2002, while the overall population of Providencia *comuna* increased (Schlack and Turnbull, 2015). For tenants, displacement is inevitably a unilateral decision in which they cannot participate.

In addition to individual private developments, two new private and public infrastructures of larger scale are planned in the area. One is a private initiative called 'Factoría Italia' that, since 2012, is building a new cultural/commercial centre restoring an old factory and theatre at the heart of the neighbourhood. This project (see Figure 10.4) will have an impact hitherto unseen in the area, bringing resources and renovation, but also many uncertainties for local residents. During the renewal process, a series of projects are taking place in sections of the former factory, as a way to use the buildings that are waiting to be transformed into the new development. These temporary uses have supposed a spectrum of activities, including cultural and artistic events, and even the installation of a presidential campaign headquarters in 2013. These temporary uses are seen by some of their promoters as an opportunity to initiate a dialogue with the community prior to the construction of Factoría Italia, seeing the temporality of these activities as an opportunity to test transformative potentials and trigger long-term processes of inclusive development (Cociña, 2013).

However, it has not been easy to start a conversation from there. A promoter of one of the temporary projects in the old factory-theatre called MilM2[4] tells the story of a conversation with an old resident in the context of complaints about loud activities in the building; the old woman, when asked about what she wanted to happen there, said that, basically, she doesn't want anything to happen: she wants everything to stay as it is.

Figure 10.4 Construction site of Factoría Italia in 2014
Source: Authors, 2014

Finally, another new public project is being built in the area, called the 'Centre of Creation Infante 1415'. For years, a public complex has hosted a health centre, public pool and old persons' home, attending mainly traditional and low-income residents of the area; using a series of properties next to it, the municipality is now enabling a new project advocating the promotion of the creative and cultural industry. According to Denise Elphick, a consultant who was in charge of the project, Infante 1415:

> is a space destined to encourage creativity, cultural production and innovation. It's a place to imagine and share; where diversity, participation and democracy are a fundamental part of its identity; it is an enabler of citizen experiences that is always under construction.
>
> Elphick, personal communication, 31 August 2015

By 2016, this project is still under construction and a series of participatory processes are taking place. What is remarkable is the potential of putting together activities that target very different visitors: both the traditional residents, who, historically, have used the public facilities, and the new ones related to creative industries, allowing a space of physical encounter and communication.

The Italia-Caupolicán case shows how, in heritage areas where commercial gentrification occurs, local identity might be needed and used as part of the creation of value – and, therefore, that there is room for more spaces of conflict resolution, in which some of the forces emerging against contested disputes can actually transform the development path of an area, participating actively in the construction of common identities.

Conclusion

In this chapter we have discussed three cases in which the existence of spaces of conflict, understood as opportunities for the encounter of clearly differentiated positions, has allowed the development of alternatives in which less affluent groups manage to remain in gentrifying neighbourhoods. Through active involvement in the process of transformation of their areas, residents have used conflict resolution in these three cases as a mechanism to exercise their right to the city, defined as the "right to change ourselves by changing the city" (Harvey, 2008: 23). These cases happen amid several and simultaneous processes of deep reconfiguration of the city, led by sometimes enormous amounts of capital invested in real estate and new commercial uses, but that at some points open room of manoeuvre for social agency and cooperation.

In all the cases presented, collective organization of residents in gentrifying areas has been key in order to leverage alternatives for inclusive development, increasing the grassroots' negotiation capacity with local governments and private sector, and enabling them to become active citizens in shaping the future of their neighbourhoods: through the development of alternative ways of housing acquisition and management in the case of IRA; through the capacity of moving a territorial discussion to the democratic arena in the case of CMSP; and through the incorporation of formal and informal instances of identity-shaping in the case of Italia-Caupolicán. In all of them, gentrification processes have implied local geopolitical transformations that have shifted the development of the neighbourhoods through contested city projects; this chapter has attempted to reflect on the underling forces behind those processes.

As Santiago is a highly segregated city, the encounter with different social classes is still seen generally as something problematic; conflict and the confrontation of clearly differentiated positions and visions are feared by many, and seen to be synonymous with violence – as something that should be avoided by all means. As we discussed in this chapter, however, the problem is that, when avoided, conflicts are usually resolved by imposing the will of those who monopolize economic and political power. In that sense, the conflicts emerging during class encounters in gentrification processes have been discussed here as opportunities to shift urban geopolitical power struggle through local mobilization of active community groups. We see, in these cases, how neoliberalism has not to be a totalizing narrative and can be, instead, contested by urban grassroots organization, triggering even cooperation among residents belonging to different social strata. The cases discussed here have shown us that collective organizations can produce conflict resolutions that allow less affluent groups to find ways to stay in gentrifying areas, opening spaces for more integrated neighbourhoods and triggering democratic processes in which differences can be reconciled.

Notes

1 We use the 2012 National Census data to present a top-line picture as it is known more disaggregated data from this census is not reliable.
2 The name of the area is contested; some neighbours will refer to it as 'Santa Isabel', while the official branding calls it 'Barrio Italia'. In this chapter we use 'Italia-Caupolicán', following the name given by Schlack and Turnbull (2015) in a recent publication.
3 MilM2 is the acronym for the Spanish for '1,000sqm', '*Mil metros cuadrados*'. The name was given because of the size of the first space in which the cultural centre operated.

References

Arriagada, C., Moreno, J. C., and Cartier, E. (2007) *Evaluación de Impacto del Subsidio de Renovación Urbana: Estudio del Área Metropolitana del Gran Santiago 1991–2006.* MINVU, Santiago.
Badiou, A. (2013) *Philosophy and the Event.* Cambridge: Polity Press.
Borsdorf, A. (2003) Cómo modelar el desarrollo y la dinámica de la ciudad latinoamericana. *EURE*, 29(86): 37–49.
Clark, E. (2005) The order and simplicity of gentrification: a political challenge. In R. Atkinson and G. Bridge (eds), *Gentrification in a Global Context: The New Urban Colonialism*. London: Routledge: 256–64.

Cociña, C. (2013) Inclusive challenges in wealthy contexts in transformation: some learnings from Barrio Italia in Santiago de Chile. In C. Boano and G. Talocci (eds), *Dpu Summerlab, 2013 Edition*, 22–23. Available from: www.bartlett.ucl.ac.uk/dpu/programmes/summerlab/summerlab_low-res.pdf. Accessed 27 August 2015.

Contreras, Y. (2011) La recuperación urbana y residencial del centro de Santiago: Nuevos habitantes, cambios socioespaciales significativos. *EURE*, 37(112): 89–113.

Contrucci, P. (2011) Vivienda en altura en zonas de renovación urbana: Desafíos para mantener su vigencia [High-rise housing in urban renewal zones: challenges for maintaining its validity]. *EURE*, 37(111): 185–9.

Cooke, B., and Kothari, U. (2001) *Participation: The New Tyranny?* London: Zed.

Corporación Barrio Italia (n.d.) *Quiénes Somos.* Available from: www.barrio-italia.cl/corporacion-barrio-italia.html. Accessed 31 August 2015. Website no longer live.

Davidson, M., and Lees, L. (2010) New-build gentrification: its histories, trajectories, and critical geographies. *Population, Space and Place*, 16(5): 395–411.

Ducci, M. E. (2000) Chile: the dark side of a successful housing policy. In J. S. Tulchin and A. Garland (eds), *Social Development in Latin America: The Politics of Reform*. Boulder, CO: Lynne Rienner: 149–73.

Fainstein, S. (1999) Can we make the cities we want? In R. A. Beauregard and S. Body-Gendrot, *The Urban Moment: Cosmopolitan Essays on the Late-20th-Century City*, London: Sage: 249–72.

Fainstein, S. (2010) *The Just City.* New York: Cornell University Press.

Freeman, L. (2006) *There Goes the 'Hood: Views of Gentrification from the Ground Up.* Philadelphia: Temple University Press.

Gilbert, A. (2002) Power, ideology and the Washington Consensus: the development and spread of Chilean housing policy. *Housing Studies*, 17(2): 305–24.

Harvey, D. (2008) The right to the city. *New Left Review*, 35: 23–40.

INE (2002) Censo Nacional de Población, Chile.

INE (2012) Resultados preliminares Censo.

Inzulza-Contardo, J. (2012) 'Latino gentrification'? Focusing on physical and socioeconomic patterns of change in Latin American inner cities. *Urban Studies*, 49(10): 2085–107.

Janoschka, M., Sequera, J. and Salinas, L. (2014) Gentrification in Spain and Latin America: a critical dialogue. *International Journal of Urban and Regional Research*, 34(4): 1234–65.

Lees, L., Shin H. B., and López-Morales, E. (2016) *Planetary Gentrification.* Cambridge: Polity.

López-Morales, E. (2011) Gentrification by ground rent dispossession: the shadows cast by large scale urban renewal in Santiago de Chile. *International Journal of Urban and Regional Research*, 35(2): 330–57.

López-Morales, E. (2013) Gentrificación en Chile: aportes conceptuales y evidencias para una discusión necesaria. *Revista de Geografía Norte Grande*, 56: 31–52.

López Morales, E. (2015) Gentrification in the Global South, *City*, 19(4): 557–66.

López-Morales, E. (2016) Gentrification in Santiago, Chile: a property-led process of dispossession and exclusion. *Urban Geography*, 37(8): 1109–31.

López-Morales, E., Gasic, I., and Meza, D. (2012) Urbanismo Pro-Empresarial en Chile: políticas y planificación de la producción residencial en altura en el pericentro del Gran Santiago. *Revista INVI*, 28(76): 75–114.

López-Morales, E., Meza, D., and Gasic, I. (2014) Neoliberalismo, regulación ad-hoc de suelo y gentrificación: el historial de la renovación urbana del sector Santa Isabel, Santiago. *Revista de geografía Norte Grande*, 58: 161–77.

López-Morales, E., Arriagada C., Gasic I., and Meza D. (2015) Efectos de la renovación urbana sobre la calidad de vida y perspectivas de relocalización residencial de habitantes centrales y peri centrales del AMGS, *EURE*, 41(124): 45–67.

Madanipour, A. (2007) Social exclusion and space. In F. Stout and R. T. LeGates (eds), *The City Reader.* London: Routledge: 158–65.

Margarit, D., and Bijit, K. (2014) Barrios y poblacion inmigrantes: el caso de la comuna de Santiago. *Revista INVI*, 29(81): 19–77.

Melillán Soto, S., and Cruz Doggenweiler, J. E. (2014) Defensa e identidad de barrio junto a la gestión Municipal en Providencia. In *Fundación Balmaceda*. Available from: http://fundacionbalmaceda.cl/defensa-e-identidad-de-barrio-junto-a-la-gestion-municipal-en-providencia/. Accessed 20 September 2015.

Miessen, M. (2010) *The Nightmare of Participation.* Berlin: Sternberg.

Milanés, P. (1980) Años [recorded by Luis Peña and Pablo Milanés] on *Años* [album].

Mouffe, C. (2005) *The Return of the Political.* London: Verso.

OECD (2012) *Working Party on Territorial Policy in Urban Areas. National Urban Policy Reviews: The Case of Chile.* Available from: www.oecd.org/officialdocuments/publicdi splaydocumentpdf/?cote=GOV/TDPC/URB(2012)13&docLanguage=En. Accessed 23 September 2014.

Renna, H. (2014) *Sobre el ejercicio y construcción de las autonomías.* Santiago: Poblar.

Rodríguez, A., and Sugranyes, A. (2004) El problema de vivienda de los 'con techo'. *Eure*, 30(91): 53–65.

Rodríguez, A., and Sugranyes, A. (2005) *Los con Techo. Un desafío para la política de vivienda social.* Santiago: Ediciones SUR.

Rojas, E. (2004) *Volver al Centro. La recuperación de áreas urbanas centrales.* Washington, DC: BID.

Sabatini, F., and Brain, I. (2008) La segregación, los guetos y la integración social urbana: mitos y claves. *EURE*, 34(103): 5–26.

Sabatini, F., Cáceres, G., and Cerda, J. (2001) Segregación residencial en las principales ciudades chilenas: Tendencias de las tres últimas décadas y posibles cursos de acción. *EURE*, 27(82): 21–42.

Sabatini, F., Robles, M., and Vásquez, H. (2009) Gentrificación sin expulsión, o la ciudad latinoamericana en una encrucijada histórica. *Revista 180. Arquitectura Arte Diseño*, (24): 18–25.

Salcedo, R. (2010) The last slum: moving from illegal settlements to subsidized home ownership in Chile. *Urban Affairs Review*, 46(1): 90–118.

Sandercock, L. (2003) City songlines: a planning imagination for the 21st century. In *Cosmopolis II: Mongrel Cities in the 21st Century.* London and New York: Continuum: 207–28.

Schlack, E., and Turnbull, N. (2009) La Colonización de barrios céntricos por artistas. *Revista 180*, 24: 2–5.

Schlack, E., and Turnbull, N. (2011) Capitalizando lugares auténticos. *Revista ARQ*, 79: 28–42.

Schlack, E., and Turnbull, N. (2015) Emerging retail gentrification in Santiago de Chile: the case of Italia-Caupolicán. In L. Lees, H. Shin and E. López-Morales (eds), *Global Gentrifications.* Bristol: Policy: 349–73.

Swyngedouw, E. (2011) *Designing the Post-Political City and the Insurgent Polis.* London: Bedford.

Vicuña del Río, M. (2013) El marco regulatorio en el contexto de la gestión empresarialista y la mercantilización del desarrollo urbano del Gran Santiago, Chile. *Revista INVI*, 28(78): 181–219.

Vigdor, J. (2002) Does gentrification harm the poor? In W. G. Gale and J. R. Pack (eds), *Brookings-Wharton Papers on Urban Affairs*: 133–73.

11 The Medellín's shifting geopolitics of informality

The Encircled Garden as a *dispositive* of civil disenfranchisement?

Catalina Ortiz and Camillo Boano

Communities do not produce city models but rather they develop life projects in their territories.

Jairo Maya, Comuna 8 Community leader, personal communication, September 2013

Medellín represents an example of urban governance innovation for Global South cities. Spectacular infrastructure of cable cars and glamorous design of public facilities located in the most conflictive areas have been the salient features of the city's transformation, heavily publicized in the city's marketing by media, international urban experts and multilateral agencies. Other core aspects of Medellín's changes less invoked in the narratives about the city involve inter-institutional coordination, long-term implementation of spatial planning tools and a public budget with exceptionally high revenues (Ortiz *et al.*, 2014), as well as aggressive security and surveillance policies. The ensemble of these myriad policies, their underlying assumptions as well as the discursive and spatial practices about the state role in dealing with informality, is what is known as the 'Medellín model' (Ortiz, 2012). An integral part of this construct is a redemption narrative of overcoming violence and inequality through a type of urbanism defined as 'social' and premised on political continuity. Despite the huge public investments and high quality architecture in informal settlements, the city remains the most inequitable city in Latin America (Abello-Colak and Guarneros-Meza, 2014; Doyle, 2016). Thus, the Medellín's case contributes to elucidate the shifting urban geopolitics of informality.

This chapter takes a critical account of the latest phase of the so-called 'Medellín model' of urban regeneration. Medellín is the symbol and the material manifestation of a successful story of a contemporary urban renaissance overcoming violence and inequality through the effective and strategic use of spatial interventions in some of the most vulnerable informal territories of the city. In particular, this chapter focuses on one of the flagship projects developed between 2011 and 2015: the Encircled Garden, whose pilot phase started in Comuna 8 – the west area of the city. This project embodied a new phase of the model that reveals not only the fragility of the political project that sustained the discourse of the 'Social Urbanism', but also the deep retrenchment of community participation. The Encircled Garden operates as the municipal response to the Metropolitan Green Belt initiative and a testing ground of the new enterprise of urban development (EDU) policy named 'Civic and Pedagogical Urbanism'. This chapter analyses how the conception and implementation of this project erodes vulnerable citizens' rights and ignores systematically their internal decision-making processes. Using Foucault's notion of *dispositive*, we argue the Encircled Garden operates as a *dispositif* of civil disenfranchisement through urban growth management and transit connectivity interventions, exacerbating the contradictions of the Medellín model.

The Encircled Garden became the linchpin strategy to tackle growth management, with the potential to become the template for addressing intervention in urban fringes. The pilot phase, carried out in Comuna 8, shows the emerging contradictions of Medellín's urban planning and design practices. The reflections in this chapter derive mainly from both authors' interest and previous research in Medellín, as well as from researching Comuna 8 as part of a series of international planning studios held by the School of Urban and Regional Studies in the National University of Colombia in Medellín (2012–14) and short summer design studios in collaboration with the Bartlett Development Planning Unit, University College London (2012 and 2013). The chapter is structured in four parts. The first section locates the idea of *dispositif* as central part of governmentality theory and its contribution to contemporary debates on informality and urban projects, elucidating the main references to Foucault's and Agamben's works. The second section provides the background of Medellín urbanization dynamics and explains why Comuna 8 and the Encircled Garden project offer a platform to explore the geopolitical meaning of the unfolding symbolic and physical implications of infrastructure provision and growth management. The third part explains how the Encircled Garden operates as a civil disenfranchisement *dispositif* through the spatial and discursive practices that enable the deprivation of dwellers' rights to remain in their territories and disavow their political priorities defined through voting. Finally, the text concludes by highlighting the shifting urban geopolitics of informality that renders planning as the design of how the life of populations is managed and controlled in contemporary social and political regimes. As a result, the Encircled Garden crystalizes the erosion of the possibilities of enacting a comprehensive set of co-produced territorial strategies to improve ecological dynamics and the living conditions of the urban fringes.

Seeing through the lens of governmentality and *dispositifs*

Foucault's governmentality renders an account of how the life of populations is managed and controlled in contemporary social and political regimes. Analysing strategic urban projects and informality through the lens of governmentality allows understanding life-administering power dedicated to inciting, reinforcing, monitoring and optimizing the forces under its control (Foucault, 2003). Foucault's governmentality is defined as an "ensemble formed by the institutions, procedures, analyses and reflections, the calculations and tactics" (Foucault, 1979: 20). In the same vein, governmentality is understood as the political rationality that explains the ways in which actors think and act in a certain way and focuses on the ways in which actors' subject the agency and shape cultural, political and socio-economic understandings (Collier, 2009; Weidner, 2010: 18; Koch, 2013). Foucault viewed it as a very specific and complex form of power that was exerted through an aggregate of physical, social and normative infrastructure – including space, architecture and its manipulation – being put into place to deal strategically with a particular problem. Therefore, governmentality includes both a managerial system of circulation in space and pastoral care for populations. Governmentality must be seen as an *ad hoc* assemblage that is introducing 'management' into diverse sites and practices in a piecemeal and contingent way in response to a dynamic and changing urban geopolitical condition. If governmentality encompasses the procedures that are designed to govern the conduct of both individuals and populations, *dispositif* "indicates the various institutional, physical and administrative mechanisms and knowledge structures, which enhance and maintain the exercise of power within the social body" (O'Farrell, 2007: 1). Even though Foucault emphasizes that power cannot be localized in a state apparatus, the state is crucial for addressing how power operates. The state is conceptualized as a "transactional reality" and part of "practices of government" (Foucault, 2003: 79–301). In other words, the state is the result of an ensemble of relations that produces the political knowledge to conduct and control populations, which operates through discourses and tools that "enable political actors to perform strategies and realize their goals" (Lemke, 2007: 48). Then, the very use of the discourse of informality constitutes the formation of a domain of intervention insofar as it operates as a central code in the reasoning style of institutions to govern populations.

The introduction of the *dispositif* analytical register in the Medellín case is far from a simple rhetorical exercise. Foucault's definition of *dispositif*, reads: "a thoroughly heterogeneous ensemble consisting of discourses, institutions, architectural forms, regulatory decisions, laws, administrative measures, scientific statements, philosophical, moral and philanthropic propositions" (Foucault, 1977: 194). Giorgio Agamben (2009) suggests elaborating on Foucault's original version, that the *dispositive*:

> is defined as a heterogeneous set of elements (discourses, regulations, institutions, architectures) and, at the same time, the network between such elements. It has a concrete strategic function and it is located in a web of power relations. Thus these are contingent relations, subject to continual change and perpetual inventiveness over time, but which produce tangible material effects – in the forms of subjectivities and in terms of specific modes of construction (of buildings, of territories and cartographies) and treatment (of people, environment, etc.).
>
> Ibid.: 11–12

In fact, Foucault's original reflections, discursive practices and governance arrangements are considered to be an aggregate of physical, social and normative infrastructure among which urban design is put in place to deal strategically with a particular problem (Boano and Talocci, 2014). As a result, the sense of politics at issue here is descriptive rather than normative and focuses on *dispositifs* that fuelled political imagination locally and globally that "penetrates the bodies of subjects, and governs their forms of life" (Agamben, 2009: 14).

That urbanism is a *dispositif* in itself is not a novelty (Secchi, 2006; Boano and Astolfo, 2015; Braun, 2014). As argued by Boano *et al.* (2013), the urban is embedded in a web of contested visions where the production of space is an inherently conflictive process, manifesting, producing and reproducing various forms of injustice, as well as alternative forces of transgressions and resistances. That is why an urban geopolitical understanding of informality and urban projects engages with the contingent nature of space production in the web of power relations. Planning policies and regulations, either holistic or selective, employ spatial devices – such as dimensions, location, separation, connection and housing typologies – that increase or decrease social difference and the distribution of welfare/wellbeing. Therefore, the focus on *dispositifs* helps to avoid a simplistic understanding of the state's use of space as object, function, or single institution to rather conceive it as a set of practices, strategies and technologies (Lemke, 2007).

Medellín, a shifting model for the urban geopolitics of informality

Medellín became a global paradigmatic model of urban renaissance after a deep crisis caused by deindustrialization, narco-trafficking and extreme urban violence. The city turned into a fertile ground for different actors and planning initiatives. It has been one of the municipalities with a long trajectory exploring planning tools. Medellín is Colombia's second largest city, situated in the region of Antioquia, north west Colombia. The city is built in the Valle de Aburrá, were the river Medellín flows. It grew as an important industrial centre, attracting waves of in-migration since the 1950s. Constrained by steep mountain slopes, informal settlements have rapidly encroached and have sprawled up the hillsides, many housing inhabitants displaced as a result of Colombia's internal conflict. Situated 1,500m above sea level it boasts of being 'the city with the eternal spring'. Medellín is divided in 16 *comunas* (districts) with a distinct north–south social divide. In this context, the city's transformation in the last decade responds to broader shifts in its governance, actively involving local government, decentralized quasi-public entities, military powers, economic elites and grassroots organizations. Medellín's local government has developed an institutional re-engineering that attempts to facilitate inter-institutional coordination, increasing public revenues and abilities to consolidate public–private alliances (Ortiz *et al.*, 2005). Moreover, the Medellín model emerges from a decisive convergence of extended practices of strategic planning, urban design and architecture that have focused local state interest and public investments in traditionally excluded peripheral neighbourhoods. These spatial interventions have concentrated on expanding the interconnected transit system through technological innovations (i.e. metro, tramway, cable cars, BRT, etc.), the generation of public spaces and the construction of multiple iconic public facilities. Notwithstanding the city's achievements and its international recognition, Medellín remains one of the most unequal Latin American cities and the territorial control of non-state armed actors still poses challenges to local governability schemes.

Medellín is shaped by its historically entrenched urban segregation, poverty and inequality. An exponential urbanization process has taken place in Medellín metropolitan area as a whole (Ortiz *et al.*, 2005). The boom of the manufacturing industry in the 1950s and 1960s attracted massive migration from rural areas and lasted a few decades, until the broader process of deindustrialization and industrial relocation hit the local economy. During the 1960s the vast majority of the self-built settlements were 'pirate neighbourhoods', where 'pirate' developers led the subdivision of rural land without infrastructure, igniting land market transactions outside the planning system, allowing residents to gradually build housing units. Simultaneously with the economic decline, during the 1970s and 1980s, Medellín played a central role in the increasing illicit international drug market. The increase in the internal armed conflict and the war on drugs in the country in late 1980s and 1990s resulted in a high peak of new settlements, created by mostly forcibly displaced inhabitants. The prevalent form of occupation was called an 'invasion', consisting of self-built housing across the steeper slopes of the Aburrá Valley, without prior land subdivision and with a different transactional mechanism than the previous 'pirate developer' (ibid.). At least 20 per cent of the urban area of the municipality of Medellín (i.e. 2,198 hectares) has an informal origin. This extension is inhabited by 48.4 per cent of the lowest-income population of Medellín (DAPM, 2005). In sum, a fifth of the urban area labelled as 'informal' by planning authorities hosts almost half of the urban population in Medellín.

Medellín had its own powerful drug cartel and suffered a traumatic violent crisis between the 1980s and the early 2000s. The rise and increasing power of the Medellín drug cartel constituted a threat to the social order. In this period, the city gained the reputation as the most dangerous of the world when the highest peak in the homicide rate reached 375 per 100,000 people in 1991. While drug lords and its army of disenfranchised youth were fighting for the hegemonic control of the territory, the war on drugs fuelled the collision with state armed groups. As a result, the war on drugs had a profound effect in the city's everyday life in these two decades. On the one hand, the constant bombings and killings imposed restrictions on mobility and fear of public life, transforming the use of the city; on the other hand, the boom of gated communities in the affluent south and the increasing self-built settlements in the fringes of the north – fuelled by forcibly displaced communities from rural areas –drastically changed the urban fabric. The deep crisis caused by de-industrialization, narco-trafficking and extreme urban violence galvanized social urgency to introduce radical changes in the city. In Colombia the introduction of a new Political Constitution in 1991 was decisive to ground legally the role of planning and the citizens engagement in the decision-making process. Iterative planning initiatives at metropolitan and local scales served as the collective catalyst to address the mid-1980 and early 1990s social crisis: from the Strategic Plan in 1996, which brought together several key actors of the city to define priorities of action; to grassroots zonal plans from 1996–9, which crystallized social mobilization initiatives; to the Municipal Territorial Plan (POT) in 1999, which synthesized more technical accounts of a spatial strategy for state regulation of land and real estate market and public works; to the integral urban projects (PUIs) that support the urban design elements of the so-called 'social urbanism' policy in 2004; to the community-led local development plans in 2007; and so forth. Hence, the cumulative effect of negotiations over priorities and planning approaches had paved the way to assemble a heterogeneous, yet disputed collective urban project. In this context, Medellín has a long history of promoting state-led interventions in informal areas to tackle the entrenched problem of violence and their habitability conditions. With the direct intervention of the national state in the early 1990s, when the city was considered the most dangerous in the world for its high homicide rate, two key programmes were implemented: Centers for Citizen Life (*Núcleos de Vida Ciudadana*) and the Programme for Integral Upgrading and Development of Informal Settlements (*Programa Integral de Mejoramiento y Desarrollo de Asentamientos Subnormales* [PRIMED]). These programmes focused on facilitating citizens' participation, community space, risk management, housing improvements and land tenure, and were the precursors of galvanizing dwellers – local state dialogue through spatial interventions. In 1999, the first POT was approved and included the delimitation of informal areas and the land management mechanisms to upgrade the selected neighbourhoods. In 2004, the local government (2004–11) coined the term 'Social Urbanism' to designate a set of state-led interventions to tackle violence and inequality, mainly through the provision of mobility infrastructure, educational public facilities and new public spaces in the crime-ridden and deprived neighbourhoods of the steeper slopes. In this period, PUIs[1] became a strategic tool to develop acupunctural interventions of public spaces and facilities in consolidated informal settlements, where the new transit infrastructure – cable cars – were introduced. A pivotal element of the recent city changes lies in the institutional realm. The function of publicly owned, decentralized institutions – run in a corporate style – has been instrumental in providing iconic architecture, public infrastructure and services. For instance, the Metro de Medellín led the construction of the only metro system in the country and has been systematically extending an integrated public transit system, with the introduction of the cable cars, tramways, connections with a bus rapid transit system and public bicycles. The introduction of two metro lines along and across the valley contributed to faster connections to the northern and southern municipalities, but planning regulations at the time disregarded the changes of land use and density along the lines. The Public Enterprise of Medellín (EPM)[2] became the strongest autonomous institution administering all public utilities provision – except telecommunications – and operating in eight countries; it owns 46 companies and delivers unparalleled revenues for the authorities to reinvest back into the city. Lastly, in the last decade, the EDU has consolidated incrementally an intervention model that encompasses the design and construction of not only public spaces and public facilities, but also the promotion of urban redevelopment and housing. Despite the efficiency and rapid service expansion and infrastructure execution of these intuitions, their accountability mechanisms are limited and, often, their interventions are disjointed from centralized planning initiatives (Ortiz and Lieber, 2014). Hence, this convergence of political alliances and institutional arrangements are at the core of Medellín's urban governance innovation.

Social Urbanism focused on accessibility and building iconic public facilities as means to gain state sovereignty in crime-prone areas. This intervention model has been scrutinized from the perspective of state transformation (ibid.); political discourses (Quinchía, 2011), participatory methods (Calderon, 2012), urban design premises (Echeverry and Orsini, 2010), the socio-spatial impact of the cable cars (Brand and Davila, 2011), their economic performance (Franz, 2016) and the relationship between urbanism and violence (Sotomayor, 2015; Samper *et al.*, 2015). Although, these works shed light on crucial factors surrounding Medellín's urban transformation, less analysis has been devoted to the political implications of the new phase of the urban projects in informal areas and the contradictions that these intervention models deploy. The following section will address the new iterations in framing state-led interventions in informal areas, particularly to tackle the occupation of its hilly fringes that has evolved informally for decades. The section will analyse how the challenges of growth management remain as some of the most salient for planning in the region and uses the case of the Encircled Garden project to reveal the disjunctions between communities needs, planning decisions and infrastructure provision is apparent in the urbanization process.

The Encircled Garden as a *dispositif* of civil disenfranchisement

Green belts are the quintessential mechanism for tackling urban growth management. Since the times of Abercrombie's green belt for London in 1935 until now, this planning tool has been reinterpreted in myriad contexts (Hack, 2012). Despite the prevalent need to regulate urban expansion processes and protect ecological areas, the conditions regarding how cities are governed and the drivers of the urbanization process have mutated enormously. Medellín's early planning initiatives in 1913 – *Plano de Medellín Futuro* and 1948 *Plan Piloto de Medellín* – have already suggested the need to contain the urban growth in the slopes of the valley (Schnitter Castellanos, 2002). However, the fast pace of the rural urban migration, the fragility of planning competences of the local state and its capacity to keep up with the requirements of the urbanization processes prevented decisive action on growth management. Nowadays, urban growth on the fringes of the city is linked to three intertwined phenomena: the spread of high-end, low-density gated communities located in the south-western high slopes of the valley driven by aesthetic consumption of landscape; the massive waves of forced displacement motivated by the internal armed conflict in the country; and the inadequate housing policy response to the provision of affordable housing and the qualitative housing deficit linked to the financiarization of the housing market and the commoditization of social housing. This complex entanglement requires reframing of the ways in which urban growth management can operate and the agency that the local state has to face those challenges. Therefore, the definition of a green belt goes beyond the delimitation of setting a barrier for new urban growth and the mechanisms to prevent new occupations emerging.

The urban growth pattern is a by-product of the migration influxes, policy regulations and imbalanced public infrastructure provision. For several decades, Medellín's planning authority considered the implementation of a green belt as a suitable territorial strategy to prevent urban expansion on the fringes. The municipality, in the last decade, embraced a two-fold strategy of growth management: that of the re-densification of central areas and containment of border expansion, as expressed in the POT. While population growth still spikes in some fringe neighbourhoods, the POT adopts the discourse of 'compact city' to advocate concentrating urban growth in the lowlands. Medellín, as the regional centre, has the imperative to address the shortage of land for new urban growth by promoting two practices: first, the protection of urban borders to help contain urban growth and, second, promoting an inward-oriented growth with emphasis on central areas near the river already equipped with excellent infrastructure. While important strategies in and of themselves, these initiatives are not being considered in tandem in order to both understand and fruitfully model their interaction with the projected transit system (Ortiz, 2014b). The implementation of a green belt was a long-standing proposal, but it was not until 2012 that the Metropolitan Planning Authority launched an initiative for developing a metropolitan green belt and the political interest from the Medellín elected Mayor, Anibal Gaviria (2012–15), pushed forward to develop a municipal version of it called the Encircled Garden. In this regard, the Encircled Garden project in Comuna 8 and its contestations emerged out of the juxtaposition of the spatial and discursive practices from the EDU, Administrative Planning Department (DAP), EPM and Metro de Medellín company, as well as the coalition of community organizations (i.e. Community Planning and Management Council, Victims Board and the Inter-Neighbourhood Board of the Disconnected [*Mesa Interbarrial de Desconectados*]). Accordingly, the Encircled Garden, in an attempt to capitalize the city legacy on urban interventions, manifests as a collision of strategies and tactics of the aforementioned actors.

Spatial narratives aim to create new imaginations but also have the potential to become empty rhetoric that justifies different spatial practices. The EDU enacted a new discursive practice to replace and update the idea of Social Urbanism and create the label 'Civic and Pedagogical Urbanism', attempting to promote conviviality and civic values in public spaces. Following that line, the EDU led the project at municipal level and described it as follows: "The Encircled Garden in Medellín is the control of expansion, housing, public space, sustainability and connection with quality of the communities that reside in the hillsides of the city. This is the flagship project of the municipality and it is characterized by being an integral

Figure 11.1 Medellín hilly landscape towards Comuna 8
Source: Boano, 2015

intervention that transforms the habitat favouring people" (Ortiz, 2014b). The Encircled Garden pilot was implemented in Comuna 8, an enclave of non-state armed groups in the central-west fringes of the city, as a strategic project that attempted to contain urban growth while delivering public space and mobility infrastructure as an extension of the metropolitan integrated transit system to connect this part of the city with downtown areas through a new tramway and cable cars. However, this interpretation of the green belt has been contested and has sparked discontent among inhabitants for obscuring other, more pressing issues for the community. In 2014, Medellín hosted the UN Habitat World Urban Forum; the Encircled Garden served as the main destination of 'urban experts' interested in understanding the benchmark for urban innovation. As a result, the mega-event became influential in legitimizing the mayor's agenda and urban marketing was instrumental in minimizing political opposition. The Encircled Garden, as the local interpretation of the metropolitan green belt, became a means by which to control particular populations. The POT, approved in 1999, set the contours of the service perimeter, defining the legal responsibility of utilities provision in the areas within. While the rate of urban expansion has decreased in absolute terms since 1999, the largest land consumption in the past decade has been focused in the high-end, low-density enclaves in the south-western fringes, outside the service perimeter (Ortiz *et al.*, 2005; Garcia, 2012). In Comuna 8[3] converge the strategic location of proximity to downtown, recent public investments in transit infrastructure, community facilities and strong community organizations that have led local planning initiatives in the last decade. Comuna 8's inhabitants are predominantly low-income families; 57 per cent of households lack land titles and make up at least 15 per cent of the total victims of internal displacement in Medellín (Samper *et al.*, 2012). Therefore, it is a clear example of how informality becomes a feature of power structures and a purposive mode of regulation, where some expressions of space production are criminalized (i.e. subaltern informalities) while other are legitimized (i.e. elite informalities) and disciplined (Roy, 2005; Hossain, 2010). Thus, the Encircled Garden becomes a crucial mechanism for targeting and trying to dissolve subaltern informalities, working in tandem with the legal prescriptions for utilities provision. It operates as an iterative set of techno-managerial decisions in the face of confrontation with local inhabitants and urban experts. Most of the success story of Medellín derives from the active role of transit in qualifying peripheries. The extension of the public transit system to the hillsides privileged enclaves of non-state armed groups during the last decade. The Metro de Medellín used the idea of 'transit as catalyst for urban transformation' and its leverage is based on its social prestige, the monopoly over the operation of public transit and technical and budgetary autonomy.

Figure 11.2 Medellín Comuna 8 local community mobilization
Source: Ortiz, 2016

In particular, the use of cable cars as public transit that adapts to the complex geography of the area constitutes the Metro company's technological innovation. From a functional perspective, the Metro company considers transit infra-' structure not only has to be built in areas that demand the services, but also that they have the capacity to generate the demand after the infrastructure is built. The technological innovation brought about a new aerial cable car system and projected to introduce tramway lines to complete an integrated metropolitan transit system. While the Metro de Medellín had, already in place, a project to connect Comuna 8 with downtown areas through the first tramway located along the Santa Elena creek and with two new cable cars lines to reach the high slopes, the EDU demanded the construction of a first phase of a monorail that could become a circuit in the high slopes without a clear linkage to the existing travel patterns. Alejandro Echeverry, champion of Social Urbanism, stated in a public forum criticizing the decision of introducing a monorail in the fragile slopes: "[The monorail] is a technology at an inadequate scale, it is economically unsustainable and gives a message about the construction on the high slopes of the city that could be a historic mistake" (El Colombiano, 2013). Whereas the previous PUI focused on informal consolidated areas, the Encircled Garden operates in the unconsolidated settlements of the fringe, where the logic of transit infrastructure as a symbol of state presence and territorial control becomes less effective. Political participation in city-making processes is an essential factor for democratic recognition. The politic organization of community-based groups in Comuna 8 is a survival strategy. The existence of strong community-led planning initiatives in Comuna 8 arises from the enduring conditions of facing non-state armed actors and the hurdles of self-construction, entangled with forced displacement. This strong community organization has promoted the Planning and Management Council (PMC) in coalition with the Board of Housing and Public Utilities, which advocates for public utilities services, and Mesa Interbarrial de Desconectados, which advocates for housing and secure tenure. In 2007, this coalition elaborated a Local Development Plan (PDL) as a legitimate bottom-up planning initiative funded by the municipality. 'Social construction of habitat' became the main discourse to mobilize Comuna 8 dwellers' interests, in an attempt to value self-built environment and stress the agency of inhabitants and advocate prioritizing integral neighbourhood upgrading (MIB). Simultaneously with the proposal of the Encircled Garden, in 2012, the municipality decided that the premises for updating the PDL in each of the 16 *comunas* of the city needed to shift. In 2007, PDLs were created based on local organization and territorially anchored NGOs focusing on enhancing inhabitants' capacities, while the new approach was to have a centralized and out-sourced process led by an international NGO. Furthermore, this turn took place in conjunction with changes in the city planning system legislation to give the responsibility of channelling citizens' demands through directly elected officials, sponsored by the municipality. As a result, the more organic leaderships and myriad community-based organizations would no longer be entitled to conduct their local planning processes.

Comuna 8 inhabitants' everyday practices of self-management, resistance and social mobilization constitute the main leverage of the PMC. The collectively crafted PDL capitalized community power/knowledge expressed in the intimate territorial experience, capabilities of community organizing and their connectedness with communities facing similar struggles in the city. Nevertheless, neither the Encircled Garden nor the construction of the tramway and cable car lines were envisioned when generating the PDL. The PMC engaged in several tactics towards the implementation of the PDL while opposing the project itself. First, in order to validate community power and knowledge, the PMC developed a partnership with a local public university (Universidad Nacional de Colombia) to co-produce a spatial narrative about Comuna 8. Second, the PMC lobbied in several participatory budgeting processes to obtain resources in order to advance risk mitigation mechanisms and public utility connections as a way to advance actions related to MIB. Third, the PMC promoted the use of a 'previous consultation', a constitutional tool that mandates a voting process before the construction of public infrastructure, for deciding on the priorities of public expenditure in the area. It was a unique use of this legal tool in an urban context and also as a self-organized initiative rather than local state-led initiative. Two local votes were held, one on 18 May 2014 and the other 13 March 2016. Both processes were important from the point of view of community-led planning: voter priorities emerged as state guarantees to remain in the territory, the reclassification of high-risk zones to lead to a mitigation plan, participatory MIB, housing with dignity, public utilities, areas for food security and inclusion of new settlements in the service perimeter. Despite the iterative process between the PMC and EDU of proposals and counter proposals, the Comuna 8 community leaders stated that:

> We denounce that the municipality is co-opting the proposals without acknowledging that they were born from the voting process, ignoring our initiatives. Still many key issues remain excluded from the project, for example the food security, the domestic public utilities provision and the security of tenure as pending themes key for the MIB.
>
> Velásquez, 2014: 87

This claim indicates that, despite the modifications of the initial project proposal, the general approach remained and, as such, deprived the urban fringe inhabitants of the right to inhabit with dignity.

The Encircled Garden was built on the faulty premise that constructing infrastructure for mobility and public space make physical barriers for new occupation. The territory of Comuna 8 is shaped by a complex overlay of myriad types of borders. For instance, the disputed administrative divisions of the neighbourhoods coexist with the invisible borders of territorial control of different non-state armed groups and the legal delimitation of the service perimeter; areas of high risk limit the state investments in the most deprived areas. For instance, the neighbourhoods of El Faro, El Pacífico, Pinares de Oriente and the sector of Alto Bonito in Villa Turbay remain outside the service perimeter (Velásquez, 2014). Furthermore, extending the logic to conceive transit infrastructure as a symbol of state presence and territorial control, the monorail is expected to act as a barrier to urban growth, controlling the territory and its local urban geopolitics. Infrastructure provision for growth management acts both as material and symbolic device in the geopolitics of urban transformation. Nevertheless, the initial conceptions of the Encircled Garden privileges the construction of a monorail, bike routes and pedestrian paths in the higher area of the Sugar Loaf hill – the border with the rural area – and neglect the provision of public utilities, risk management works and neighbourhood upgrading. In order to limit urban expansion, the Encircled Garden paradoxically uses the monumentalization of the mobility infrastructure to gain sovereignty and securitization and ignores risky areas: inclusion is the visible face of territorial control. Access to water and sanitation are basic rights. Nowadays, about 3,000 housing units located in high-risk areas lack sanitation and about 1,000 housing units lack potable water (ibid.). The EPM has the legal duty to provide utilities – particularly water and sanitation – to households located within the service perimeter and outside areas considered as high risk in the POT. A significant proportion of the current urban population has struggled not only to build their own dwellings, but also to gain rights to provision of basic utilities and services. Since the first settlements established in the area during the 1970s, water provision and circulation paths were mostly self-built by the dwellers as a community leader suggests:

> The neighbourhoods in the mountain slopes were made with the hands of the neighbours through the union, human warmth, the joint venture… where the municipal administration did not participate decisively in our lives. That is why it is necessary to acknowledge the relevance of the community organization beyond the requests to a deaf state that is insensitive and useless when it comes to a guarantee to those who build these neighbourhoods, this city and its richness.
>
> Mesa Interbarrial de Desconectados, 2014

The expansion of the utilities system by the EPM took several years to arrive, but only got as far as the areas closer to the downtown area because these higher-slope settlements were considered to be outside the perimeter of services, failing to recognize a lived, albeit self-made settlement. The EPM's new approach to service provision in low-income areas is introducing pre-paid water and electricity to allow self-monitoring of energy consumption and prevent the cutting off of the service, as was a common practice with households that were in debt. As a reaction to this practice, the Board of Utilities and Housing highlighted that:

> we defend the public and communitarian management of the water not for profit and follow the principles of social and solidarity economy… We refuse to implement pre-paid potable water because it limits the right to have water as a common good, in a city where 46,000 households live without potable water.
>
> Ibid.

This perspective problematizes the EPM assumptions about considering citizens as customers and the company strategies as a way to avoid lack of payment and cumulative debt in the utilities charges instead of the social investment of the high profits the company receives.

Housing and safety are essential rights. The biopolitics of being classified as high risk results in preventing not only utilities provision, but also in being disqualified from applying for subsidies for housing upgrading. The recently approved POT (2014) found that, in Comuna 8, the quantitative housing deficit is at least 8,000 units (DAP, 2014), highlighting the pressing issues not tackled directly by the Encircled Garden. The very techno-political definition of risk and the delimitation criteria of high-risk areas have changed; while the 2006 POT determined that Comuna 8 had 7,682 constructions at high risk, the updated plan in 2014 identified that only 1,708 constructions were at high risk (Samper *et al.*, 2015). As a reaction to the proposal of the Encircled Garden, the PMC leaders declared:

> The priority is risk reduction rather than building monorails. For us the true path to build a life with dignity is to bring back what the Programme for Integral Upgrading and Development of Informal Settlements – PRIMED – did in the 1990s where the risk management was paramount to improve the housing, the legalization and land titling processes within a proposal of 'Neighbourhood upgrading'.
>
> Mesa Interbarrial de Desconectados, 2014

The common claims contained in the plan refer to economic development, housing affordability, risk mitigation, utilities provision and secure land tenure. In reaction to the social mobilization, the EDU drafted a housing programme called Sustainable Neighbourhoods; this programme got to the phase of technical studies and design without certified analysis about micro-zoning risk assessment, yet eviction threats remain. According to the public pronouncement of the community leaders:

> Here, the municipality cannot come and show us a map with red areas that are at risk; we need the studies that support those areas. In other words, we need technical backing to say that we need to leave but, so far, there is not a technical argument.
>
> Mesa Interbarrial de Desconectados, 2014: 2

Therefore, Encircled Garden is, in fact, a distortion of their priorities, needs and desires because it privileges a monorail, bike routes and pedestrian paths in the higher area of the Sugar Loaf hill –and postpones public utilities, risk management works and neighbourhood upgrading.

Conclusion

We have argued that the Encircled Garden operates as a disenfranchisement *dispositf*. The case of Comuna 8 reveals the contradictions of state-led spatial interventions. The convergence of a strategic location, recent public investment in community facilities and a strong community organization make Comuna 8 a unique city sector in terms of strategies of transformation. The PMC has been able to position its agenda while expressing its discontent with local government for ignoring pressing issues and community priorities. Furthermore, in a context governed by the rhetoric of participatory planning, mobilization and dissensus are often criminalized. In sum, the conception and implementation of the Encircled Garden project has threatened several citizen rights: (1) the right to a safe life, by ignoring the environmental risk of the households located in the fragile slopes; (2) the right to remain, brought about by the evictions and the uncertainty in the housing removals; (3) the right to participate, through the erosion of community-led planning and the out-sourcing of the local development plans. Here, the *dispositif* at play in the urban geopolitical landscape encapsulates the trajectories of the struggles in the production of urban spaces and urban infrastructures, but also describes planning regulations in their discursive, spatial and symbolic operation. In this context, informality turns into a "heuristic device that uncovers the ever-shifting urban relationship between the legal and illegal, legitimate and illegitimate, authorized and unauthorized" (Roy, 2011: 233). The merit of the *dispositive* – as developed by Foucault and further developed by Agamben – is that it enables us to grasp both the various geopolitical *forms* that administration and government takes today and the provisional and *ad hoc* nature of the knowledge, practices and institutions that are stitched together into a heterogeneous totality. Informality – and the geopolitics that are generated to tame, control and deal with its inscriptions in urban dynamics – is the manifestation of our contemporary form of biopolitics. We use the case of Comuna 8 illustrate where state-led interventions monumentalized mobility infrastructure to regain territorial control over people, spaces and forms of life. In the context of the Encircled Garden project, we explored the unfolding aggregate of physical, social and normative infrastructures depicted as a key socio-technical process that shapes urban space – that rather than focusing on political representation, sanitation, security tenure, housing adequacy and risk mitigation, emphasizes the iconicity of infrastructure and architecture as instrumental to the political ends of symbolic conquest of unruly territories and as a tool of urban geopolitical control.

Regional comparative reflections increased the status of Medellín as paradigmatic example in the urban geopolitics discourse of Latin America, as well as in global discussion around local government leadership in urban transformations. We were able to diagnose interactions between political agendas, architecturally invasive branded projects, architectural ego and urban marketing influencing a renewed urban discourse. The spectacular imagery of the these interventions has transformed how "a critical area of the city was perceived by insiders and outsiders... leading to relevant social, socio-spatial and socio-economic revitalization, while promoting inclusive patterns of urbanization" (Blanco and Kobayashi, 2009: 76) – becoming another example of urban design that, masked with social discourses, capitulates to neoliberal urbanization and state-led control, losing the opportunity to close the circle between abstraction and representation and the site specificity of architecture. Medellín's hyperbolic transformations – infused with urban marketing, the co-option of social movements by the rhetoric of urban equity, accessibility and governability – could be seen as a complex set of practices, both discursive and material, that include the essence of the city and its material composition into a perfectly representable new urban subject. But being a 'subject' in the Foucauldian framework is not simply being subjected and assigned to subordination, but is also to develop a set of resistances and reactive forces that oppose such normative definitions.

Notes

1 The first and most salient PUI was built in Comuna 1, including a cable car line [Line K] and Santo Domingo Library Park and the PUI in Comuna 13 with another cable car line [Line J], electric escalators and the San Javier Library Park.
2 EPM started its operation in 1955 and its plans of sewerage and aqueduct network expansion have been crucial for directing the growth patterns of the metropolitan area, responding to an increasing demand and the limitations of the geography.
3 In 2005, Comuna 8 had more than 156,706 inhabitants distributed in 18 neighbourhoods and 32 sectors in 577 hectares. More than 35,000 inhabitants are located in the urban fringes and about 25,000 are forcedly displaced by the armed conflict (Comuna 8 Local Development Plan [PDL], 2015).

References

Abello-Colak, A., and Guarneros-Meza, V. (2014) The role of criminal actors in local governance. *Urban Studies*, 51(1): 1–22.

Agamben, G. (2009) *What is an Apparatus?* Stanford, CA: Stanford University Press.

Benjamin, W. (2005) Fragment 74: capitalism as religion. In E. Mendieta (ed.), *Religion as Critique: The Frankfurt School's Critique of Religion*. New York: Routledge.

Blanco, C., and Kobayashi, H. (2009) Urban transformation in slum districts through public space generation and cable transportation at north-eastern area: Medellín, Colombia. *Journal of International Social Research*, 2(8): 75–90.

Boano, C. (2014) Architecture of engagement: informal urbanism and design ethics. *Atlantis Magazine*, 24(4): 24–8.

Boano, C., and Astolfo, G. (2015) Speculations on the Italian rhetoric of mending peripheries. *Quaderns: The Journal of the Association of Architects of Catalonia*. Available from: http://quaderns.coac.net/2015/01/mending-peripheries/. Accessed 22 March 2017.

Boano, C., and Talocci, G. (2014) Fences and profanations: questioning the sacredness of urban design. *Journal of Urban Design*, 19(5): 700–21.

Boano, C., Hunter, W., and Newton, C. (2013) *Contested Urbanism in Dharavi: Wrigins and Projects for the Resilient City*. London: Development Planning Unit.

Brand, P. (2011) Governing inequality in the South through the Barcelona model: 'social urbanism', in Medellín, Colombia. Interrogating urban crisis: governance, contestation, critique. 9–11 September, De Montfort University. Available from: www.dmu.ac.uk/documents/business-and-law-documents/research/lgru/peterbrand.pdf. Accessed 21 June 2016.

Brand, P., and Davila, J. (2011) Mobility innovation at the urban margins, Medellin's metrocables. *City*, 15(6): 647–61.

Braun, B. (2014) A new urban dispositif? Governing life in an age of climate change. *Environment and Planning D: Society and Space 2014*, 32: 49–64.

Calderon, C. (2012) Social urbanism: participatory urban upgrading in Medellin, Colombia. In L. Yildiz and P. Kellett (eds), *Requalifying the Built Environment: Challenges and Responses*. Göttingen: Hogrefe. Available from: file:///C:/Users/Tucker /Downloads/Calderon%202012-Social%20Urbanism.pdf. Accessed 21 June 2016.

Collier, S. (2009) Topologies of power: Foucault's analysis of political government beyond 'Governmentality'. *Theory Culture Society*, 26(6): 78–108.

DAPM (Departamento Administrativo de Planeacion) (2005) Documento Tecnico de Soporte de la revision del Plan de Ordenamiento Territorial de Medellin. Alcaldia de Medellin.

DAP (Departamento Administrativo de Planeacion) (2014) Documento Tecnico de Soporte de la revision del Plan de Ordenamiento Territorial de Medellin. Alcaldia de Medellin.

Davila, J. (ed.) (2013) *Urban Mobility and Poverty: Lessons from Medellín and Soacha, Colombia*. London: DPU, UCL and Universidad Nacional de Colombia.

Debord, G. (1992) *Society of the Spectacle*. London: Rebel.

Doyle, C. (2016) Explaining patterns of urban violence in Medellin, Colombia. *Laws*, 5(3): 2–17.

Echeverry, A., and Orsini, F. (2010) *'Informalidad y Urbanismo Social en Medellín'. Informality and Social Urban Planning in Medellin. Sostenible?* Available from: http:// upcommons.upc.edu/revistes/bitstream/2099/11900/1/111103_RS3_AEcheverri_%20 P%2011-24.pdf. Accessed on 28 September 2013.

El Colombiano (2013) Arzobispo de Medellín cuestiona situación actual de la ciudad. 1 March. Available from: http://www.elcolombiano.com/historico/arzobispo_de_medellin_ cuestiona_situacion_actual_de_la_ciudad-KEEC_231281. Accessed 21 June 2016.

Fabricius, D. (2008) Resisting representation: the informal geographies of Rio de Janiero. *Harvard Design Magazine*, 28, Spring/Summer: 1–8.

Foucault, M. (1977) The confession of the flesh (1977) interview. In C. Gordon (ed.), *Power/Knowledge Selected Interviews and Other Writings*, 1980: 194–228.

Foucault, M. (2003) *Abnormal Lectures at the College de France 1974–1975*. London: Verso.

Foucault, M. (2008) *The Birth of Biopower: Lectures at the college du France 1978–1979*. London: Palgrave.

Frantz, T. (2016) Urban governance and economic development in Medellín: an 'urban miracle'? *Latin American Perspectives*, 43(5): 1–16.

Garcia, J.-H. (2012) *Slum Tourism and City Branding in Medellin*. Available from: https:// blog.inpolis.com/2013/01/28/guest-article-slum-tourism-and-city-branding-in-medellin/. Accessed 4 December 2016.

Hack, G. (2012) Shaping urban forms. In B. Sanyal, L. J. Vale and C. D. Rosan (eds), *Planning Ideas that Matter: Livability, Territoriality, Governance, and Reflective Practice*. Cambridge, MA: MIT Press: 33–64.

Hossain, S. (2010) Informal dynamics of a public utility: rationality of the scene behind a screen. *Habitat International*, 35(2): 275–85.

Koch, N. (2013) 'Spatial socialization': understanding the state effect geographically. *Nordia Geographical Publications*, 44(4): 29–35.

Lemke, T. (2007) An indigestible meal? Foucault, governmentality and state theory. *Distinktion: Scandinavian Journal of Social Theory*, 8(2): 43–64.

Maclean, K. (2015) *Social Urbanism and the Politics of Violence: The Medellin Miracle*. Basingstoke: Palgrave Macmillan.

Mesa Interbarrial de Desconectados (2014) Comunicado a la opinión pública sobre el proceso de Mejoramiento Integral de Barrios (MIB) en la Comuna 8 de Medellín. Available from: www.ciudadcomuna.org/ciudadcomuna/images/noticias/2014/julio/Comunicado%20MIB%20Comuna%208.pdf. Accessed 16 December 2016.

O'Farrell, C. (2007) *Key Concepts*. Available from: www.michel-foucault.com/concepts/. Accessed 4 December 2016.

Ortiz, C. (2012) Bargaining space: deal-making strategies for large-scale renewal projects in Colombian cities. PhD thesis, unpublished document. Available from: http://indigo.uic.edu/handle/10027/9152. Accessed 22 March 2017.

Ortiz, C. (2014a) Negotiating conflicting spatial narratives in Medellin. In G. Talocci and C. Boano (eds), *SUMMERLAB 2013*. London: University College London. Available from: www.bartlett.ucl.ac.uk/dpu/programmes/summerlab/summerlab_low-res.pdf. Accessed 21 June 2016.

Ortiz, C. (2014b) Social urbanism in Medellin as 'hyper-upgrading': caveats on a widely circulating 'best practice'. Presented at Latin American Studies Association Conference at Chicago, USA. Assembling the Contemporary Latin American City: South-South Circuits, Planning Exchanges and Policy Mobilities.

Ortiz, C. (2015) Designing from the cracks: The potentials of the Medellin Model drawbacks, in G. Talocci, C. Boano, eds., SUMMERLAB 2014, Available from: www.bartlett.ucl.ac.uk/dpu/programmes/summerlab/DPU_summerLab_Pamphlet_2014. Accessed 21 June 2016.

Ortiz, C., and Lieber, L. (2014) Medellín: Hacia la construcción de un modelo de estrategias para la equidad territorial? In *Equidad Territorial en Medellín: La Empresa de Desarrollo Urbano como motor de la transformación urbana*. InterAmerican Development Bank: 30–60.

Ortiz, C. *et al.* (2005) Medellin: dynamics of metropolitan growth. In *Colombia: Land and Housing for the Urban Poor*. University College of London, Colombian National Department of Planning and Cities Alliance Programme.

Ortiz, C., Navarrete, J., and Donovan, M. (2014) Reflexiones sobre la equidad territorial: de la concepción a la acción, in *Equidad Territorial en Medellín: La Empresa de Desarrollo Urbano como motor de la transformación urbana*. InterAmerican Development Bank: 12–29.

Quinchía, S. (2011) Discursos y prácticas de planeación: el caso del urbanismo social de Medellín, Colombia. Unpublished MA dissertation, School of Urban and Regional Planning, Universidad Nacional de Colombia, Medellín.

Rao, V. (2006) Slum as theory: mega-cities and urban models. In C. G. Crysler, S. Cairns and H. Heyden (eds), *The SAGE Handbook of Architectural Theory*. London: Sage: 671–86.

Roy, A. (2005) Urban informality: toward an epistemology of planning. *APA Journal*, 71(2): 147–51.

Roy, A. (2011) Slumdog cities: rethinking subaltern urbanism. *International Journal of Urban and Regional Research*, 35(2): 223–38.

Saldarriaga, J.-L. (2013) El monorriel puede ser un error histórico: académicos. Available from: www.elcolombiano.com/historico/el_monorriel_puede_ser_un_error_historico_academicos-MCEC_255988. Accessed 2 December 2016.

Samper, J., Ortiz, C., and Soto, J. (2015) *Rethinking Informality: Strategies of Urban Space Co-Production*. Cambridge, MA: School of Architecture and Planning,

Massachusetts Institute of Technology. 232pp. Available from: http://issuu.com/jo tasamper9/docs/medellin_workshop_en_3-26-15_web1. Accessed 21 June 2016.

Schnitter Castellanos, P. (2002) José Luis Sert y Colombia. De la Carta de Atenas a una Carta del Hábitat. Doctoral thesis. Fernando Alvarez Prozorovich, Escuela Técnica Superior de Arquitectura, Universidad Politécnica de Cataluña, Barcelona, June: 389.

Secchi, B. (2006) The rich and the poor, comment vivre (ou ne pas vivre) ensemble. In P. Viganò and P. Pellegrini, *Comment vivre ensemble*. Budapest: Officina: 374.

Sotomayor, L. (2015) Equitable planning through territories of exception: the contours of Medellin's urban development projects. *International Development Planning Review*, 37(4): 373–97.

Velásquez, C. (2014) Comuna 8 Aprobó Mediante Consulta Popular Su Propuesta Por El Derecho A Vivir Dignamente En Nuestros Territorios. *Revista Kavilando*, 6(1): 80–7. Available from: http://www.ssoar.info/ssoar/handle/document/43843. Accessed 21 June 2016.

Weidner, J. R. (2010) Globalizing governmentality: sites of neoliberal assemblage in the Americas. *FIU Electronic Theses and Dissertations*, 258. Available from: http://digital-commons.fiu.edu/cgi/viewcontent.cgi?article=1319&context=etd. Accessed 21 March 2017.

12 Assessing critical urban geopolitics in Foz do Iguaçu, Brazil

Peter D. A. Wood

Introduction

Cities throughout Latin America demonstrate important contexts for diversity, inequality and participation in the globalizing world. From Iberian colonialism to contemporary patterns of migration, populations in Latin American cities have instilled and been subjected to a variety of political systems and ruling classes. Many of these systems have evoked considerable contestation and disdain. Recent protests in Venezuela (BBC, 2014), Brazil (*The Economist*, 2013; Romero, 2015), Chile (Taylor, 2011), Argentina (Goni and Watts, 2012) and Ecuador (Alvaro, 2015), to name key examples, remind us how tenuous and tense internal relations in Latin American countries have been. Specifically regarding urban issues, systems of geopolitical control and hegemony are often seen as causal factors in relation to the nature of cities (Graham, 2004a). Cities are characterized as the locations where geopolitical action takes material form, places reflecting the costs and risks of global expansion and connectivity. Rather than recognizing cities as origins of geopolitical power themselves, the urban scale is often attributed the passive role of an immediately visible result of action taken at other scales. In many ways, this is a useful frame of analysis for capturing the under-represented views of development and the nexus between urban and geopolitical regimes.

Marginality is a central theme of development and urbanism, one that crosses disciplines and world regions (Moser, 1989; Perlman, 1979; Wacquant, 2008). Despite this, studies of urban geopolitics often fail to account for the multidirectional relationship between urbanism and geopolitics. Rather than simply canvases upon which outside actors (multinational corporations, national governments, international trade blocs) paint pictures of development and the effects of globalized capital, cities complicate and even challenge the authority vested within these traditional forms of geopolitics. In this chapter I use a borderland in Latin America to test for divergence and overlap in perspective among development stakeholders. Borderlands provide examples of frequent migration, national security and trade (Brunet-Jailley, 2007; Lindquist, 2009; Walker, 1999). In Latin America, borders help exemplify the attempts and challenges to controlling regional and global flows which come as a result to systems of geopolitical rule and influence. Attention to Latin American border metropolises is important, as studies of urban marginality and development tend to focus on internationally recognized municipalities (Garmany, 2011). Metropolitan areas located within borderlands offer particular insight towards how the urban and geopolitical influence one another. Incorporating Southern perspectives on growth in these areas adds to the Global North-dominated frameworks of urban studies theory (Robinson, 2011; Roy, 2009). Adding data on large cities outside areas like Rio de Janeiro, São Paulo, Buenos Aires and Bogotá, as well as the continental coastline, similarly helps broaden the purview of urban geopolitics and the circumstances within which geopolitics is experienced and executed. Expanding the areas of study additionally enriches research on contested cities by providing more diverse settings in which contestation takes form.

Background theory: urban geopolitics, critical geopolitics

The recent turn of geopolitics towards the urban is the latest in a series of adaptations to classic geopolitics (Agnew, 2003; Dalby, 1990, 1994; Tesfahuney, 1998). Since the 1970s, four central lines of thought have been explored within geopolitics: realist, political economy, critical and feminist (Purcell, 2006). Each of these addresses issues of power hierarchies and experiences of global connectivity in varying lights. As a result, inequality and security have received much attention from scholars of geopolitics since the emergence of a critical geopolitical eye. The rise of these themes is partly attributable to their trans-scalar nature: security and inequality are often by-products of national and international agendas, yet are visibly experienced on a personal, daily level. Much of geopolitics has become interdisciplinary, with various theoretical and empirical backgrounds contributing to the geopolitical study (Dodds, 2001). Through this interdisciplinary work, the incorporation of critical analysis has affected geopolitics as an area of study. The rise of critical geopolitics has brought attention to the role 'everyday' geopolitics has within studies of territorialized power (Pain, 2009; Pain and Smith, 2012). Rather than limiting analysis to official policies by government leaders, additional sources of geopolitical influence have been given scholarly visibility and helped shape this 'critical turn'. What is particular about this trend is how urban phenomena have received little attention as points of focus for critical geopolitical research. While calls have recently been made for increased study of urban geopolitics, these calls do not contain an explicitly critical element (Yacobi, 2009). Urban areas are, instead, discussed as settings upon which deep power relations are projected. Through exploration of urban marginality, the need for an inverse mode of study (in which cities are more than 'projected upon') becomes more apparent.

When discussing urban geopolitics the concepts of slums and war zones, in particular, become prominent as visualizations of how exploitation, inequality and marginalization assume form (Graham, 2004b, 2006). Slums often embody processes of resource extraction, exploitation and neoliberal strategies to both enhance global connectivity and protect these paths of connection from perceived threats. War zones similarly reflect a result of geopolitical struggle expressed through militaristic means. If brought specifically to a Latin American context, urban geopolitics commonly involves discussion of *favelas, villas* and *barrios* (terms among the varied regional nomenclature), though within urban geography the points of focus studied for understanding these phenomena have shifted in recent years (Knopp and Kujawa, 2006). Rather than shift away from slums and related issues of marginality, scholars and non-academics involved with geopolitics can engage with these spaces as more expansive representations of networks of power, influence and the resulting scales of those excluded from social authority, also termed 'subaltern actors'. Within Latin America, studies of urban geopolitics can (and ought to) move towards revealing structural barriers to development and integration. This will lead to addressing similarities and differences of the views development stakeholders maintain, views that affect the implementation of development plans at the urban and global scales. Though issues of marginality and security remain key areas of study in the geopolitics of Latin America, scholars and practitioners must move towards a more expansive understanding of how urbanism affects and is affected by geopolitical processes. An expanded understanding of urbanism is essential for growing regional studies of Latin America and broadening the domain of geopolitical enquiry.

Subalternity through an urban lens: critical urban geopolitics

Synthesizing critical and urban geopolitics results in more than simply a critical view of geopolitics in an urban context (Brenner *et al.*, 2012). Moving these sub-disciplines together draws attention to the need for understanding and studying geopolitics critically from a range of scales. In this sense, geopolitics follows the definition from Ó'Tuathail and Agnew as a form of discourse that moves beyond descriptive measures of foreign policy and towards more holistic understandings of how power is constructed and exerted (cited in Dodds, 2001). The urban scale provides an important and useful starting point for its accessibility and reflection of both smaller- and larger-scale processes. An urban vantage point can lead to a more nuanced exploration of systems of power as they relate to other scales of experience. Subaltern experience, a crucial component to studying such systems of power, tends to be presented in certain forms within urban and geopolitical studies. In the former, subalternity is characterized as corresponding to those of low socio-economic status, especially people of colour (Gidwani, 2006). In the latter, subaltern geopolitics, subalternity is discussed as a challenge or subversion of state-centred power by key groups or individuals (Sharp, 2011; Sidaway, 2012). Even when critically studied as a matter of both the urban and world regions, subalternity typically refers to a person or place, and rarely an ideology or perspective (Bayat, 2000). Testing for variance in opinion and ideology contributes to a fuller understanding of the power dynamics between marginalized populations and those with more authority in an area. This, in turn, can help to identify who is associated with under-represented viewpoints and what their limited participation in the public process signifies.

Participation in urban planning is often associated with the idea of addressing inequality and including marginalized populations (Miraftab and Kudva, 2014). It is less common for participation to focus on representing a wide range of ideologies or worldviews. Even when minority viewpoints are brought into discussion, they are often incorporated as tokens used to instill a false sense of diversity (Arnstein, 1969). The idea of including seemingly marginalized groups in development efforts is not new (Booth, 1979; Gittell and Shtob, 1980), though the action often bears little on-the-ground significance (Miraftab, 2009). Yet moving beyond this, to the point of basing inclusionary practice on perspectives rather than simply demographics, is rarely discussed explicitly as an objective for development planning. Harnessing urban geopolitical analysis helps bring clarity to the ways various scales of experience contribute to and are affected by urbanism and global connectivity.

Of the many contexts displaying power struggles, urban border regions are a useful example of how urban and regional development are entry points for understanding changes in national form. The observation has been made before, with Brenner (1999, 2000, 2004) bringing attention to the utility of cities as newly relevant hubs of policy and action. This analysis, nonetheless, focuses more on domestic policy and economics, and less on the transnational, geopolitical components of urbanization, growth and planning. In this sense, the need for critical analysis of urban–geopolitical nexuses – especially in non-European settings – becomes visible. Using participation in urban development planning as a frame of analysis reveals how scholarship like that of Brenner can be aptly expanded upon with the inclusion of more regions in the Global South. Focusing on Southern cities brings new places of study into view, the Brazilian *favela* being a common example of a hard-to-replicate regionally specific phenomenon. Some of this research has been done under the category of megacities (Badshah and Perlman, 1996; Castells, 1998; Kraas, 2007), though such scholarship arguably recreates dichotomies between North and South as central and peripheral spaces and puts disproportionate focus on select 'global' cities (Garmany, 2011; Parnell and Robinson, 2012; Watson, 2009). Yet, just as favelas and other slums are cited as evidence of uneven development producing polarized experiences and spaces, those same *favelas* need to be seen as more than dystopian wastelands and end products of systems of global capital. Brazil is a particular example of what Amar (2009) calls a new form of war, one pitting the marginalized urban poor against state-sponsored policing. The processes affecting global flows of capital and migration can be generalized across space, but regional peculiarities necessitate a more nuanced approach to understanding the ways in which urban participation takes form. In this sense, both Northern and Southern subaltern urban experiences can inform one another (Schindler, 2014).

When expanding beyond subalternity in the Global South and moving towards trans-border integration of marginalized populations in Latin American cities, contributing scholarship begins to dwindle. Much cross-border integration has been focused on European localities, and not Latin America or the Global South (Church and Reid, 1996; Kratke, 1998; Matthiessen, 2005), though exceptions do apply (Shen, 2004; Yang, 2005). In the Americas, NAFTA and the US–Mexico border are the closest example of research on Latin American border integration (Fernández-Kelly and Massey, 2007; Herzog, 1991), although there are, again, occasional exceptions (Castillo, 2003; Martínez, 1998). Here I present findings from a Latin American border region known for connecting a wide range of people, places and ideologies. Both the juxtaposition of multiple countries and the location of this particular border in relation to trade blocs and international trading partners contribute to the high connectivity seen in the area. Understanding perspectives on power dynamics in this Latin American borderland is important for moving urban geopolitics into new literal and figurative territory.

Critical urban geopolitics in Foz do Iguaçu

Observing opinions can help bring attention to under-represented views and populations in a study area. In order to measure opinions on development partici-pation, a Q method study (see below for explanation) was conducted in Foz do Iguaçu, Brazil, and Ciudad del Este, Paraguay, during late 2014. This region is often referred to as the *triple frontier* or *tri-border area* (Hudson, 2003; Madani, 2002). The three cities comprising the triple frontier are Foz do Iguaçu (estimated 2015 population 263,782 [IBGE, 2015]), Ciudad del Este (estimated 2015 popu-lation 290,912 [Dirección General de Estadística, Encuestas, y Censos, 2015]) and Puerto Iguazú, Argentina (2010 population 41,062 [Instituto Provincial de Estadística y Censos de Misiones, 2013]). The area is known for several key characteristics: it is a logistically important border city in the heart of Mercosur (the Common Market of the South); the largest hydroelectric dam by volume in the world (Itaipu Binational) is located there; the world-renowned Iguaçu Falls are present; and there is a high level of ethnic diversity in the metropolitan region. This latter point of diversity has been particularly relevant for issues of interna-tional security. Following two 1990s bombings of Jewish centres in Argentina and the attacks of September 11, 2001, the large Arab population of Foz and Ciudad del Este led foreign officials to suspect that political Islamist terror groups were gathering funding from illicit and informal sales across the Brazilian–Paraguayan border (Abbott, 2004; Kittner, 2007; Levitt, 2008; Mendel, 2002; Sverdlick, 2005). As these accusations have become less common, locally based projects aimed at cross-border urban integration and development have gained momentum. On the Brazilian side of the border, Codefoz (Economic and Social Development Council of Foz do Iguaçu) was initiated in November 2012. In 2014, Codeleste (Economic and Social Development Council of Ciudad del Este) began formation in the Paraguayan city. Codespi (Economic and Social Development Council of Puerto Iguazú) shortly thereafter began the organiza-tional process in Argentina. Aside from these three co-aligned organizations, Foz do Iguaçu leadership has focused on progressing two main projects: Beira Foz (an effort to attract private sector investment and development on the margins of the Paraná River, which divides Brazil and Paraguay) and a second bridge to comple-ment the Friendship Bridge between Foz and Ciudad del Este crossing the Paraná River (see Figure 12.1).

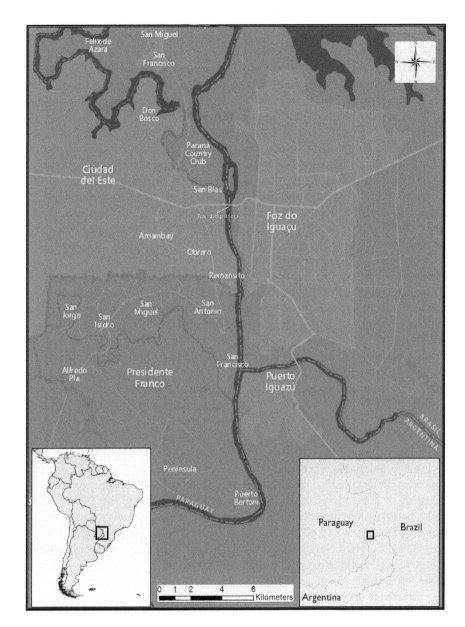

Figure 12.1 Foz do Iguaçu metropolitan area
Source: Openstreetmap, n.d.

Methods and fieldwork in Foz do Iguaçu

In order to better understand the scales and structures of participation within an internationally significant urban area, I collected Q sorts from 30 stakeholders in Foz do Iguaçu development. Q method originated in psychology and has become a widely used methodology for measuring perspectives among a target population (Watts and Stenner, 2005). In Q studies, participants arrange a predetermined set of opinionated statements, ranking them between extremes (in this case 'strongly disagree' to 'strongly agree') on a semi-normalized curve. This method was chosen because of its utility in assessing what general views exist among participating individuals. Rather than compare answers to specific survey questions, Q quantifies the relationship between opinionated statements to delineate distinguishable points of view. In this project, I chose to measure what narratives were prominent regarding public participation in Foz do Iguaçu development. Doing this helps demonstrate what priorities exist in this border region and how these perspectives compare and contrast with others.

When observing and recording discourse on urban geopolitics it is important to recognize the various types of experiences recognized by involved actors. Through Q method, a condensed set of self-referential views on a topic is produced (Durning, 1999; van Eeten, 2001). As Brown (1993) points out, only a finite amount of viewpoints exists on a given subject before most views become exhausted. For this reason, Q studies do not require a large sample size, nor are they necessarily improved through continuous addition of participants (Barry and Proops, 1999; van Exel and de Graaf, 2005). Upon review, results of the Q study are used to critically engage with urban geopolitics: do urban experiences suggest top-down geopolitics is reciprocated outward from the city? Answering this question, while noting other significant findings, applies urban geopolitical study in an under-represented region while using development participation as an entry point for understanding deeper power relations.

Prior to fieldwork, I expected sentiment against US and European corporate and governmental influence to shape perspectives. Considering post-colonial research on Latin American exploitation and subjugation to the Global North (and, by extension, neoliberal economic policy), presumed resistance to outside control of resources shaped the hypothesis for this project (Mignolo, 2005). From this view, I expected participants to favour autonomy of development planning to avoid further exploitation of the region. Because the triple frontier so vividly juxtaposes urban inequality with international security between countries, I expected any differences in opinion to correlate with national ties.

In order to collect Q statements to be sorted by participants, I conducted seven semi-structured interviews, attended a local environmentalist meeting, an economic development community meeting and reviewed news stories published online, academic journal articles and a video lecture series from a conference organized by the Foz-based *100 Fronteiras* magazine. This phase of data collection took place in August and September 2014. From these materials, I derived over 200 statements used as a foundation for the study (called a concourse in

Q methodology). An advantage to including interview data and similar qualitative sources (rather than creating concourse statements on one's own) is that using outside statements helps capture an organic realism to the issues being researched. This also helps preserve meaning and intent when dealing with participants whose native language differs from that of the researcher. To develop a concourse of statements along these lines I had to determine what types of expressed opinions I was seeking. Through formulation based upon grounded theory (Charmaz and Belgrave, 2012), I asked questions aimed at sparking expansive discussion on development in the Foz do Iguaçu area, what projects and investment opportunities existed or should exist, what long-term ambitions development should work towards, what obstacles prevented development from progressing smoothly and what the state of public participation in the development process was in this tri-national region. Beyond interview data, I extracted statements from publicly available documents and videos.

After collecting a sufficient number of statements to form a concourse, I then selected 36 statements to be used in Q sorting. Rather than use a survey to collect data, often considered problematic when used to gauge public opinion, I chose to use Q method to give participants more control over the study matter (Bertrand and Mullainathan, 2001). In my study, I ultimately selected six categories of statements: (1) security, (2) national/geopolitical-level concerns, (3) commerce, (4) municipal planning projects, (5) regional planning/collaboration and (6) participation/representation in planning. Within these six groupings, statements addressed multiple scales, themes and points of opinion. Using six categories, each containing six statements, allowed for an even distribution of sorted material within a semi-normal distribution. The purpose of this study, and any Q project, is to analyse the relationship between multiple fragments of information as they form a broader fabric of understanding. In this chapter, data collected from 30 stakeholders in the Foz do Iguaçu, Brazil, development (27 Brazilians, three Paraguayans: 21 males, nine females) is used to identify prevalent views in the area.

Case study results and interpretation

Using PQ Method software, all results were analysed and participants were categorized by similar responses. Of the 30 participants, 29 fell into three categories of overarching viewpoints, with one participant remaining unclassified for lack of statistically significant correlation with any factor. These results, summarized below, compare the three factors by z-score (shown as F1, F2 and F3, representing all three factors) for each of the 36 statements used. Negative z-scores correlate with disagreement, while positive scores reflect agreement with the statement. The 36 statements used were presented in Spanish or Portuguese to participants. All quotations are from post-sort interviews with participants, translated by the author.

Participants showed varied responses to the 36 sorted statements. The role of national ties was less prominent than expected. This is partly because of the limited participation of non-Brazilians, and also because national ties were less influential than views on which nation (if any) is a suitable leader for the region.

Factor 1: local integration optimists

> The border has a dynamic that is different from the capitals.
>
> Development coordinator, 19 September 2014,
> in Ciudad del Este, Paraguay

Seventeen participants were categorized together and labelled local integration optimists (15 Brazilians, two Paraguayans). People in this group were likely to have positive outlooks regarding international cooperation between Brazil, Paraguay and Argentina. These individuals were also likely to see city-level projects as an ideal platform for this integration to grow. Overall, they supported integrative security efforts, and saw projects like the second bridge between Paraguay and Brazil as examples of necessary economic connectivity. These respondents also were likely to view small businesses as important institutions who suffer as a result of federal taxes and policies.

The highest scoring statement for this group, *Integrated security between Foz do Iguaçu, Ciudad del Este and Puerto Iguazú is still needed*, reflects the common sentiment that trans-border unity at the urban level is the best starting point for addressing geopolitical threats. The lowest scoring statement, *Foz do Iguaçu is unable to make a project encompassing international will*, demonstrates how individuals in this category maintain optimistic outlooks regarding the collaboration of urban and national interests. This also suggests that those in this category disagree with the notion that geopolitical change necessarily comes from national capitals or megacities such as Asunción or São Paulo. Regarding national ties, this category reflected common views across national identities that equality of cities and urban-controlled development plans are prioritized.

Factor 2: institution sceptics

> Their [Codefoz] interest is the interest of a small group of people... the biggest number of people from that system of Codefoz, for them [participation] makes no difference. In reality it hinders. If they were actually public, big business doesn't win.
>
> Environmental NGO director, 18 September 2014,
> Foz do Iguaçu, Brazil

Seven participants were categorized and labelled as institution sceptics (all from Brazil). These participants were likely to express scepticism and cynicism regarding projects aimed at integrating across national boundaries, particularly when these projects were created by politicians. A prominent example of this attitude is disapproval of Mercosur, an example of multiple countries working together for economic gains. In addition, this group commonly expressed that structural and systemic features of the triple frontier region made corruption and instability likely to arise. Those categorized under this factor also tended to perceive a lack of centralized organization and planning in Foz do Iguaçu. Compared to F1, people in this group were less likely to support bi- or tri-national integration through development projects.

F2's highest-scored statement, *The tri-border area between Argentina, Brazil and Paraguay offers a refuge that is geographically, socially, economically and politically very conducive to permitting the activities of organized crime, Islamic terrorist groups and corrupt workers*, demonstrates the cynicism and distrust these individuals hold towards many influential institutions. The lowest-scored statement, *Mercosur is extremely positive for Foz do Iguaçu because there is effective integration*, confirms the claim that those in this category view the Mercosur regional trade bloc unfavourable in relation to Foz development. Such strong disagreement also corresponds to the broader view from this group that projects like Mercosur and Codefoz, though seemingly participative and open, are façades used to further the influence of those in power. When considering the regional tendency for considering corruption common in Latin American politics, the emergence of this factor is unsurprising (Canache and Allison, 2005). Only Brazilian citizens were assigned to this grouping. A possible explanation is that the sorted statements focused on development in the Brazilian city of Foz; those in Paraguay were less likely to be familiar with Foz municipal politics or have clear opinions that would lead to explicitly sceptical views.

Factor 3: nationalists

> Foz doesn't depend on its neighbours, but the opposite: the neighbours depend on Foz.
>
> Alderman, 19 September 2014, Ciudad del Este, Paraguay

Five participants were categorized and labelled as nationalists, whose views reflect support for Brazilian-led efforts in Foz do Iguaçu development (four Brazilians, one Paraguayan). Individuals from this group were likely to view the Brazilian federal government as a natural leader for growth and development in the border region. Those in this group felt strongly that illicit trans-border activity had moved northward, rather than stopping, following several years of increased border control in the area. This was often accompanied by the perspective that Mercosur is a detrimental institution for growth and progress in the triple frontier region. Compared to F1, nationalists were more likely to dismiss Paraguayan presence in planning efforts and to maintain Brazilian national-level institutions as the origin of desirable change in the border region. Compared with F2, nationalists were less likely to view the triple frontier as conducive to fostering corruption, crime and terrorism, and more likely to view Paraguay and Argentina as less developed than Brazil. This group's views generally reflect similar perspectives of Brazil becoming a rising regional and global power (Dauvergne and Farias, 2012; Reid 2014).

This group's highest scoring statement, *Today it is much easier to ship cocaine, marijuana and guns across the border between Paraguay and Mato Grosso do Sul than in another city*, relates to the general perception that Brazilian security remains under threat, despite attempts to control cross-border traffic near Foz do Iguaçu. The border crossing referenced in the statement lies north of the triple frontier; that those in this factor feel trafficking has moved there (where the land border is harder to monitor than the bridge between Foz and Ciudad del Este) suggests they feel Brazil and Paraguay are fundamentally different. The lowest-scored statement, *Mercosur is extremely positive for Foz do Iguaçu because there is effective integration*, corresponds to the right-of-centre criticisms this group has of the left-leaning trade bloc. When seen as a detriment to development in Brazil – and, by extension, Foz – Mercosur is taken to be a highly unfavourable economic union. The inclusion of one Paraguayan participant in this category is significant: favouring nationalist views does not necessarily entail the assumption that one's home country is the preferable national power for regional leadership. In this case, Brazil is held as superior even by one participant who serves as alderman in Ciudad del Este, Paraguay.

When analysing the factors by statement category, the three factors reflect the following general perspectives, as shown in Table 12.1.

Further analysis shows the variance in opinion between stakeholders on particular issues. Overall disagreement between participants was best encapsulated by statements on international collaboration and perceived inequality among the three bordering countries. Participants were most likely, in comparison, to express similar sentiment on statements regarding Mercosur and small businesses.

Follow-up survey results

After collecting all 30 sorts and conducting Q factor analysis, all participants were contacted via email with their factor name and a brief description. Participants had the option to return a short questionnaire asking if they agreed with their assigned category, which other stakeholders they expected to agree or disagree with and if Paraguayans, Argentinians and groups in Foz receive disproportionate representation or lack thereof in Foz development participation. Five participants returned questionnaires; the five ranked their agreement with the assigned factor (1 being 'strongly disagree', 10 'strongly agree') 8/10, 8/10, 9/10, 9/10 and 4/10. Most free-form responses were brief, though one participant from F2 opined, "The civil community is not participatory. There are not representatives and groups with effective and participatory leaders" (environmental NGO director, personal email, 20 December 2014).

Table 12.1 Factor summaries by issue

Statement category	Factor 1: local integration optimists	Factor 2: institution sceptics	Factor 3: nationalists
Security	Border security is a multinational issue in which cities have a duty to collaborate. Criminals are adaptive to physical and political environments, thus integration is essential for preventing gaps in policy or practice.	The porosity of the border reflects deeper problems of government accountability. In order to curtail trafficking, municipal integration must be better formed.	Federal presence is needed to impede illicit trans-border activity. Smugglers and traffickers are opportunistic, therefore controlling their movement is challenging.
International inequalities	Differences in economies and cultures exist, though working across borders can reduce the negative externalities of high inequality. Regional integration is useful if it establishes common priorities among those involved.	The creation of a regional trade bloc must be evaluated critically. Differences in wealth and livelihood between countries tend to be exaggerated.	Brazil is economically advanced within South America. Growth and development in Brazil benefits its neighbours just as greatly.
Commerce	Through lower federal taxes and free trade, economic growth can occur at the city level. With a stronger economy, poverty and crime are less likely to occur.	Trans-border trade is often imbalanced and exploitative. Neither urban nor federal institutions can improve economic standing on their own.	Domestic production is favoured over importation. Brazilian businesses are a preferable leader for improved relations with neighbours.
Urban development	Removing barriers to cross-border movement is important for developing a city's economy. Enhanced connectivity between cities is in the interest of all municipalities and their residents.	The state of planning at the urban level is poorly organized. Rather than focus on expansion, development planning needs to emphasize improvement of underutilized areas in the urban core.	Unified, transparent municipal organization is needed for sustainable growth. What benefits Brazilian urban growth also benefits its neighbouring cities.
Regional integration	Progress benefiting multiple bordering countries should start with urban-level institutions. Development on the border needs to prioritize the metropolitan region over individual cities.	Current forms of regional trade integration are inadequate. In order to improve circumstances at the urban-level, institutions at other scales need to assume equal responsibility.	The trade bloc, in its current status, insufficiently integrates the region. Brazilian interests need to be prioritized over concerns of neighbouring countries.
Participation and representation	Border cities possess different characteristics to national capitals and globally connected megacities. Locally driven planning can sufficiently represent diverse sectors of the border region while striving to spur growth from the urban scale outward.	Representing multiple scales of governance is vital yet very delicate. Neither federal nor municipal agencies sufficiently foster diverse participation.	National institutions have a central role in improving urban livelihood. Municipal efforts to integrate multiple voices in development planning fall short of their objective.

Source: Author, 2015

Conclusion

Past work contributing to urban geopolitics suggests that terror, security and fear shape urban spaces (Pain, 2010; Pain *et al.*, 2010). The results of this study suggest a slightly different picture: urban spaces may also affect systems of security, fear and power, as well as the ways they are perceived. Urbanism and geopolitics share a multidirectional relationship. As cities advance within changing geopolitical and geoeconomic environments, generating policy which accounts for this relationship can help mitigate social uprisings, reshape cities and reconfigure the networks of power which produce inequality and uneven urban geopolitical participation. The shared border region of Foz do Iguaçu demonstrates the delicate balance between national and municipal interests within development projects in the twenty-first century amid an era of rapid global connectivity. Markedly different national economies, dissimilar capabilities for capital accumulation and mutual recognition for border security (executed in different ways) are characteristics exacerbated in this frontier region. All active parties are not equally represented in development discussions, however, necessitating an examination of who is empowered, why and what consequences come as a result of this imbalance.

While some participants view outside influence as prevalent at the triple frontier, outside control was more commonly perceived as coming from Brasilia, Asunción, or Buenos Aires. This revealed how defining 'outsider' or 'foreigner' is a complicated process in Latin America, especially in a border city like Foz do Iguaçu. Those critical of 'equal participation' as a selling point for urban-driven development planning tended to perceive projects like urban development council Codefoz to be falsely labelled as openly participatory. This criticism is partly explained by unequal standards of living across national borders. Differences in per capita wealth certainly exist between Brazil, Paraguay and Argentina ($11,384, $4,712 and $12,509, respectively, for 2014 in 2016 US$ [The World Bank, 2016]) – differences which suggest that truly 'equal' participation is difficult or impossible to attain. In contrast, those statements that reflected consensus among participants suggest that small businesses and locally controlled commerce are generally favourable. This is peculiar in Foz, since many of the development projects are funded or organized by the binational Itaipu dam, which is jointly owned and operated by the Paraguayan and Brazilian governments.

Current discussions over participation in Latin American development, whether along the border or elsewhere, inevitably deal with Mercosur and the challenges of regional integration (e.g. Grimson and Kessler, 2014; Filho and Rückert, 2013). Dynamics within this trade bloc demonstrate the complexity of development strategies and how they are received in the region. Inter- and intra-trade union inequality of members is an important aspect to economic development plans (e.g. Blyde, 2006; Hinojosa-Ojeda, 2003; Frenkel and Trauth, 1997; Dunford, 1994); by laying a theoretical foundation, these discrepancies can be understood more clearly. Differences exist both across borders and within Foz do Iguaçu, demonstrating how geopolitics at the urban scale is often divided between elite control and subaltern disempowerment. By acknowledging overlap and incompatibilities between views of stakeholders, those studying urban geopolitical participation can better consult municipal authorities on the challenges and opportunities at hand. In doing this, urban geopolitics can simultaneously grow more critical in nature, more influential in application and be better prepared to address issues of contested cities as they arise globally.

References

Abbott, P. K. (2004) Terrorist threat in the tri-border area: myth or reality? *Military Review*, 84(5): 51–5.

Agnew, J. A. (2003) *Geopolitics: Re-visioning World Politics*, 2nd edn. London: Routledge.

Alvaro, M. (2015) Protesters in Ecuador demonstrate against Correa's policies. *The Wall Street Journal*, 26 June. Available from: www.wsj.com/articles/protesters-in-ecuador -demonstrate-against-correas-policies-1435279037. Accessed 10 January 2016.

Amar, P. (2009) Operation Princess in Rio de Janeiro: policing 'sex trafficking', strengthening worker citizenship, and the urban geopolitics of security in Brazil. *Security Dialogue*, 40(4–5): 513–41.

Arnstein, S. R. (1969) A ladder of citizen participation. *Journal of the American Institute of Planners*, 35(4): 216–24.

Badshah, A. A., and Perlman, J. E. (1996) Mega-cities and the urban future. *City*, 1(3–4): 122–32.

Barry, J., and Proops, J. (1999) Seeking sustainability discourses with Q methodology. *Ecological Economics*, 28(3): 337–45.

Bayat, A. (2000) From 'dangerous classes' to 'quiet rebels' politics of the urban subaltern in the Global South. *International Sociology*, 15(3): 533–57.

BBC (2014) What lies behind the protests in Venezuela? Available from: www.bbc.com/ news/world-latin-america-26335287. Accessed 10 January 2016.

Bertrand, M., and Mullainathan, S. (2001) Do people mean what they say? Implications for subjective survey data. *The American Economic Review*, 91(2): 67–72.

Blyde, J. (2006) Convergence dynamics in Mercosur. *Journal of Economic Integration*, 21(4): 784–815.

Booth, J. A. (1979) Political participation in Latin America: levels, structure, context, concentration and rationality. *Latin American Research Review*, 14(3): 29–60.

Brenner, N. (1999) Globalisation as reterritorialisation: the re-scaling of urban governance in the European Union. *Urban Studies*, 36(3): 431–51.

Brenner, N. (2000) The urban question as a scale question: reflections on Henri Lefebvre, urban theory and the politics of scale. *International Journal of Urban and Regional Research*, 24(2): 361–78.

Brenner, N. (2004) *New State Spaces: Urban Governance and the Rescaling of Statehood*. New York: Oxford University Press.

Brenner, N., Marcuse, P., and Mayer, M. (2012). *Cities for People, Not for Profit: Critical Urban Theory and the Right to the City*. London: Routledge.

Brown, S. R. (1993) A primer on Q methodology. *Operant Subjectivity*, 16(3/4): 91–138.

Brunet-Jailly, E. (2007) *Borderlands: Comparing Border Security in North America and Europe*. Ottawa: University of Ottawa Press.

Canache, D., and Allison, M. E. (2005) Perceptions of political corruption in Latin American democracies. *Latin American Politics and Society*, 47(3): 91–111.

Castells, M. (1998) Why the megacities focus? Megacities in the new world disorder. *Publication MCP-018, Mega-Cities Project, New York.* Available from: http://mega-cities.net/pdf/publications_pdf_mcp018intro.pdf. Accessed 22 October 2014.

Castillo, M. Á. (2003) Mexico–Guatemala border: new controls on transborder migrations in view of recent integration schemes? *Frontera Norte*, 15(29): 35–65.

Charmaz, K., and Belgrave, L. L. (2012) 'Qualitative interviewing and grounded theory analysis'. In Jaber F. Gubrium (ed.), *The SAGE Handbook of Interview Research: The Complexity of the Craft.* Thousand Oaks, CA: Sage: 347–66.

Church, A., and Reid, P. (1996) Urban power, international networks and competition: the example of cross-border cooperation. *Urban Studies*, 33(8): 1297–318.

Dalby, S. (1990) American security discourse: the persistence of geopolitics. *Political Geography Quarterly*, 9(2): 171–88.

Dalby, S. (1994) Gender and critical geopolitics: reading security discourse in the new world disorder. *Environment and Planning D*, 12(5): 595–612.

Dauvergne, P., and Farias, D. B. (2012) The rise of Brazil as a global development power. *Third World Quarterly*, 33(5): 903–17.

Dirección General de Estadística, Encuestas, y Censos (2015) *Proyección de la Población por sexo y edad según Distrito, 2000–2025 – Revisión 2015.* Available from: www.dgeec.gov.py/Publicaciones/Biblioteca/proyeccion%20nacional/Proyeccion%20Distrital.pdf. Accessed 15 January 2016.

Dodds, K. (2001) Political geography III: critical geopolitics after ten years. *Progress in Human Geography*, 25(3): 469–84.

Dunford, M. (1994) Winners and losers: the new map of economic inequality in the European Union. *European Urban and Regional Studies*, 1(2): 95–114.

Durning, D. (1999) The transition from traditional to postpositivist policy analysis: a role for Q-methodology. *Journal of Policy Analysis and Management*, 18(3): 389–410.

Fernández-Kelly, P., and Massey, D. S. (2007) Borders for whom? The role of NAFTA in Mexico–US migration. *The ANNALS of the American Academy of Political and Social Science*, 610(1): 98–118.

Filho, C. P. C., and Rückert, A. A. (2013) Estratégias de cooperação e desenvolvimento nas fronteiras do MERCOSUL: a Região Transfronteiriça do Iguaçu. *Anais: Encontros Nacionais da ANPUR*, 15. Available from: www.anpur.org.br/revista/rbeur/index.php/anais/article/view/4322. Accessed 23 June 2014.

Frenkel, M., and Trauth, T. (1997) Growth effects of integration among unequal countries. *Global Finance Journal*, 8(1): 113–28.

Garmany, J. (2011) Situating Fortaleza: urban space and uneven development in northeastern Brazil. *Cities*, 28(1): 45–52.

Geenhuizen, M. V., van der Knaap, B., and Nijkamp, P. (1996) Trans-border European networking: shifts in corporate strategy? *European Planning Studies*, 4(6): 671–82.

Gidwani, V. K. (2006) Subaltern cosmopolitanism as politics. *Antipode*, 38(1): 7–21.

Gittell, M., and Shtob, T. (1980) Changing women's roles in political volunteerism and reform of the city. *Signs*, 5(3): S67–S78.

Goni, U., and Watts, J. (2012) Argentina protests: up to half a million rally against Fernández de Kirchner. *Guardian*, 9 November. Available from: www.theguardian.com/world/2012/nov/09/argentiana-protests-rally-fernandez-kirchner. Accessed 10 January 2016.

Graham, S. (2004a) *Cities, War, and Terrorism: Towards an Urban Geopolitics*. Malden, MA: Blackwell.

Graham, S. (2004b) Postmortem city: towards an urban geopolitics. *City*, 8(2): 165–96.

Graham, S. (2006) Cities and the 'War on Terror'. *International Journal of Urban and Regional Research*, 30(2): 55–276.

Grimson, A., and Kessler, G. (2014) *On Argentina and the Southern Cone: Neoliberalism and National Imaginations*. New York: Routledge.

Hansen, N. (1983) International cooperation in border regions: an overview and research agenda. *International Regional Science Review*, 8(3): 255–70.

Herzog, L. A. (1991) Cross-national urban structure in the era of global cities: the US–Mexico transfrontier metropolis. *Urban Studies*, 28(4): 519–33.

Hinojosa-Ojeda, R. A. (2003) Regional integration among the unequal: a CGE model of US–CAFTA, NAFTA and the Central American Common Market. University of California, Los Angeles. Available from: www.cepal.org.mx/www2/sica/ESTUDIOS/RH/Informe%20final.pdf. Accessed 23 June 2014.

Hudson, R. (2003) *Terrorist and Organized Crime Groups in the Tri-Border Area (TBA) of South America*. Federal Research Division. Washington, DC: Library of Congress.

IBGE (Instituto Brasileiro de Geografia e Estatística) (2015) *Diretoria de Pesquisas – DPE – Coordenação de População e Indicadores Socias – COPIS*. Available from: http://cidades.ibge.gov.br/xtras/perfil.php?codmun=410830. Accessed 15 January 2016.

Instituto Provincial de Estadística y Censos de Misiones (2013) *Gran Atlas de Misiones*. Posadas, Argentina: IPEC.

Kittner, C. C. B. (2007) The role of safe havens in Islamist terrorism. *Terrorism and Political Violence*, 19(3): 307–29.

Knopp, L., and Kujawa, R. (2006) Urban geography. In Barney Warf (ed.), *Encyclopedia of Human Geography*. Thousand Oaks, CA: Sage: 517–22.

Kraas, F. (2007) Megacities and global change: key priorities. *The Geographical Journal*, 173(1): 79–82.

Kratke, S. (1998) Problems of cross-border regional integration: the case of the German–Polish border area. *European Urban and Regional Studies*, 5(3): 249–62.

Levitt, M. (2008) *Hamas: Politics, Charity, and Terrorism in the Service of Jihad*. New Haven: Yale University Press.

Lindquist, J. A. (2009) *The Anxieties of Mobility: Migration and Tourism in the Indonesian Borderlands*. Honolulu: University of Hawaii Press.

Madani, B. (2002) Hezbollah's global finance network: the triple frontier. *Middle East Intelligence Bulletin*, 4(1). Available from: www.meforum.org/meib/articles/0201_12.htm. Accessed January 2016.

Martínez, M. V. (1998) El suroeste de Venezuela: espacios de integración fronteriza. *Anales de Geografía de la Universidad Complutense*, 18: 139–58.

Matthiessen, C. W. (2005) The Öresund area: pre- and post-bridge cross-border functional integration: the bi-national regional question. *Geo Journal*, 61(1): 31–9.

Mendel, W. W. (2002) Paraguay's Ciudad del Este and the new centers of gravity. *Military Review*, 82(2): 51.

Mignolo, W. D. (2005) *The Idea of Latin America*. Malden, MA: Blackwell.

Miraftab, F. (2009) Insurgent planning: situating radical planning in the Global South. *Planning Theory*, 8(1): 32–50.

Miraftab, F., and Kudva, N. (2014) *Cities of the Global South Reader*. Hoboken: Taylor and Francis.

Moser, C. (1989) Gender planning in the Third World: meeting practical and strategic planning needs. Gender and Planning Working Paper 11. Development Planning Unit, London.

Openstreetmap (n.d.) Available from: api06.dev.openstreetmap.org/. Accessed 25 January 2017.

Pain, R. (2009) Globalized fear? Towards an emotional geopolitics. *Progress in Human Geography*, 33(4): 466–86.

Pain, R. (2010) The new geopolitics of fear. *Geography Compass*, 4(3): 226–40.

Pain, D. R., and Smith, P. S. J. (eds) (2012) *Fear: Critical Geopolitics and Everyday Life*. Aldershot: Ashgate.

Pain, R., Panelli, R., Kindon, S., and Little, J. (2010) Moments in everyday/distant geopolitics: young people's fears and hopes. *Geoforum*, 41(6): 972–82.

Parnell, S., and Robinson, J. (2012). (Re)theorizing cities from the Global South: looking beyond neoliberalism. *Urban Geography*, 33(4): 593–617.

Perlman, J. (1979) *The Myth of Marginality: Urban Poverty and Politics in Rio de Janeiro*. Berkeley: University of California Press.

Purcell, D. (2006) 'Geopolitics', in Barney Warf (ed.), *Encyclopedia of Human Geography*. Thousand Oaks, CA: Sage: 184–6.

Reid, M. (2014) *Brazil: The Troubled Rise of a Global Power*. New Haven, CT: Yale University Press.

Robinson, J. (2011) Cities in a world of cities: the comparative gesture. *International Journal of Urban and Regional Research*, 35(1): 1–23.

Romero, S. (2015) Protests across Brazil raise pressure on President Dilma Rousseff. *New York Times*, 16 August. Available from: www.nytimes.com/2015/08/17/world/americas/brazilians-protest-to-urge-president-dilma-rousseffs-ouster.html. Accessed 10 January 2016.

Roy, A. (2009) Strangely familiar: planning and the worlds of insurgence and informality. *Planning Theory*, 8(1): 7–12.

Schindler, S. (2014) Understanding urban processes in Flint, Michigan: approaching 'subaltern urbanism' inductively. *International Journal of Urban and Regional Research*, 38(3): 791–804.

Sharp, J. (2011) A subaltern critical geopolitics of the war on terror: postcolonial security in Tanzania. *Geoforum*, 42(3): 297–305.

Shen, J. (2004) Cross-border urban governance in Hong Kong: the role of state in a globalizing city-region. *The Professional Geographer*, 56(4): 530–43.

Sidaway, J. D. (2012) Subaltern geopolitics: Libya in the mirror of Europe. *The Geographical Journal*, 178(4): 296–301.

Sverdlick, A. R. (2005) Terrorists and organized crime entrepreneurs in the 'triple frontier' among Argentina, Brazil, and Paraguay. *Trends in Organized Crime*, 9(2): 84–93.

Taylor, A. (2011) Student protests in Chile. *The Atlantic*, 10 August. Available from: www.theatlantic.com/infocus/2011/08/student-protests-in-chile/100125/. Accessed 10 January 2016.

Tesfahuney, M. (1998) Mobility, racism and geopolitics. *Political Geography*, 17(5): 499–515.

The Economist (2013) The streets erupt. *The Economist*, 18 June. Available from: www.economist.com/blogs/americasview/2013/06/protests-brazil. Accessed 10 January 2016.

van Eeten, M. J. G. (2001) Recasting intractable policy issues: the wider implications of the Netherlands civil aviation controversy. *Journal of Policy Analysis and Management*, 20(3): 391–414.

van Exel, J., and de Graaf, G. (2005) *Q Methodology: A Sneak Preview*. Available from: http://citeseerx.ist.psu.edu/viewdoc/download?doi=10.1.1.558.9521&rep=rep1&type= pdf. Accessed 21 March 2017.

Wacquant, L. (2008) *Urban Outcasts: A Comparative Sociology of Advanced Marginality*. Cambridge: Polity Press.

Walker, A. (1999) *The Legend of the Golden Boat: Regulation, Trade and Traders in the Borderlands of Laos, Thailand, China, and Burma*. Honolulu: University of Hawaii Press.

Watson, V. (2009). Seeing from the South: refocusing urban planning on the globe's central urban issues. *Urban Studies*, 46(11): 2259–75.

Watts, S., and Stenner, P. (2005) Doing Q methodology: theory, method and interpretation. *Qualitative Research in Psychology*, 2(1): 67–91.

World Bank (2016) GDP per capita. *World Development Indicators*. Available from: http:// data.worldbank.org/indicator/NY.GDP.PCAP.CD. Accessed 15 January 2016.

Yacobi, H. (2009) Towards urban geopolitics. *Geopolitics*, 14(3): 576–81.

Yang, C. (2005) Multilevel governance in the cross-boundary region of Hong Kong: Pearl River Delta, China. *Environment and Planning A*, 37(12): 2147–68.

Part V

Comparative discussion

Jonathan Rokem

This book's main intention is to bring *geopolitics* into the mainstream of *urban studies* to enhance our understanding of cities as contested nexus points of social, spatial and political change. In light of this overarching aim, we critically discuss in this concluding dialogue the comparative turn in urban studies, beyond the call for *non-Western* post-colonial theory formation. This concluding fifth part is not meant to serve as a comprehensive review of the diverse and unique individual chapters in the book. Each case speaks for its own manifestations of how planning is framed within local urban geopolitics, drawing lessons for the rich world of cities, within a growing relational interest to learn from across different geographical territories and political contexts.

In a series of conversations about the book's theoretical and geographical scope, the book editors Jonathan Rokem and Camillo Boano reflect on several topics related to the very nature of geopolitical urbanism in the contemporary global system with Michael Safier. A key founding figure of the Development Planning Unit, University College London, Michael shares from his vast personal interest in *cosmopolitan urbanism* and *planning* as an overarching approach to make sense of the changing global forces, and the prerequisite to reframing what has been labelled as the *post-colonial* in relation to *urban geopolitics* and *planning* in a world of *contested cities*. This is followed by a concluding note from one of the prominent scholars in political geography James D. Sidaway, raising some profound theoretical questions about the nature and scope of the 'urban' and 'geopolitics'.

13 Geopolitics, cosmopolitanism and planning

Contested cities in a global context

Michael Safier with Jonathan Rokem and Camillo Boano

This concluding piece stems from the book editors Jonathan Rokem (JR) and Camillo Boano's (CB) conversations in 2015–16, with Michael Safier (MS). Michael Safier, an economist and geographer, joined the Development Planning Unit (DPU) at its 1971 inception in the University College London (UCL). Since 1990, he has been developing ideas about cosmopolitanism, cosmopolitan development and cosmopolitan urbanism in relation to planned intervention in the context of contested and divided cities in a global perspective. In a set of conversations about the book's theoretical and geographical approach, Michael shared with us his personal interest in *cosmopolitan urbanism* as an overarching approach to make sense of changing global forces, and the prerequisite to reframing what has been labelled as the *post-colonial* in relation to *urban geopolitics* and *planning* in different contested cities.

JR: How can we clarify the status of the 'post-colonial' as a predominant but perplexing concept in a current world order that has passed through a period of decolonization, producing a world of new nation states?

MS: The idea of the post-colonial has produced a whole subject discipline in the historical and social sciences that is paralleled by other post-discourses, of which the most celebrated have been the postmodern, the post-structural and the post-developmental. I argue that the nature of the term suggests that we have gone beyond a substantive concept of the colonial, the modern and the developmental, but are still umbilically linked to these discourses. The issue here is how to articulate a new order with a new substance, akin to the still debated notions of 'new world orders' that supposedly advance our conception of geography and politics to deal with the shifts in global power and influence that have clearly come to pass in recent decades. I argue, further, that a more immediate concern needs to be given to illuminating not the post-colonial but the emergence of a new colonialism, the 'neo-colonial', that reproduces in a new form the relation between new colonial powers and new colonial territories, where asymmetric power relations are again evident, as between new imperialists and new subalterns, congruent with the contemporary world order.

CB: Can you provide some good examples of these arguments?

MS: Neo-colonial practices are widely present in the range of interventions that are meant to implement the corpus of progressive international relations that have been promoted since the Second World War and its aftermath between 1942 and 1948. The historical trajectory of decolonization has been paralleled by the complementary trajectory of universal human rights, conflict resolution and humanitarian intervention, the practices which were meant to be undertaken by and on behalf of the 'international community', principally represented by the United Nations through its various programmes covering dispute resolution, peace building and the protection of collective rights of civilian communities in times of conflict. These ambitions have been prominently articulated in the resolutions and practices embodied in the 'Responsibility to Protect' doctrine passed by the UN in 2005. In principle, these advances are intended to establish a just and equitable relation between those nations who need any and all kinds of assistance beyond their own capacity to provide, and those nations with the power to provide that assistance on behalf of an international community that embraces both parties in a context of mutual respect. In reality, the record of international interventions over the past 40 years has consistently failed to follow, in whole or part, any such universal programme. Even when nominally under the authority of the United Nations, interventions have been carried out by agents of individual nation states that have had, and still have, the predominance of power and resources applied across the world to advance human rights, respond to humanitarian emergencies and resolve violent conflicts. Individual nation states still operate, first and foremost, in their own national interests and subordinate the needs and demands of other nations, especially so-called 'failed' or more 'fragile' states and communities with which they interact – in the context of huge disparities between major powers such as the USA, Russia, China and India, Brazil and Indonesia, and the majority of small and weak states most often afflicted by conflict and extreme conditions. Neo-colonial relations mimic the colonial, most crucially in the range of war and post-war interventions, which are often invasions akin to super-size 'gunboat diplomacy', as in Afghanistan, Iraq and Libya; and as in subsequent hegemonic rule, as with the colonial governor-like position of the 'High Representative' in post-war Bosnia, supervising constitutional arrangements and ordering political and administrative reform following on from the Dayton peace agreement of 1995.

JR: How far do we need to de-centre the still current predominant unit of account that is the sovereign nation state, and to dismantle its current constitutive power over all other state agencies?

MS: The international nation-state system is so strongly embedded in both power and consciousness that it presents a highly problematic reality as the constitutive unit in the current world order. The degeneration of domestic politics and government, and of effective action over non-state actors, in many fragile or compromised small states – in the era of 'new wars' crossing both round and within national boundaries, and with the greatly expanded sphere of regional and global 'social movements' in civil society forming transnational networks – requires rethinking the Westphalian order. It is increasingly the case that the city has become and is becoming a rising unit of account in world affairs, but has yet to be given the comparative accounting frameworks that are still used exclusively for nation-state assessments. The notion of the nation state as the natural container of freedom, democracy and cultural community, the inheritance of Enlightenment liberalism, has more recently been deformed by a simplified majoritarian vision: one state representing one tomography, one territory, one authority and one identity is increasingly at odds with the fluidity, mobility and intersection allegory of the contemporary world.

CB: What would be a way forward in conceptualizing a new global territorial order?

MS: Over the last half-century there have been many simplified, often binary categorizations of world geography and economic/political order: North, South, East, West; first, second and third worlds; the West and the rest; and unipolar, bipolar and increasingly multipolar designations of power relations. These are all massive, oversimplified and indiscriminate notions that I have referred to as elements of a 'lumpen' intellectual-cum-popular discourse. In contrast, a fundamental component of cosmopolitan thinking has been to advance the epistemic position of 'as well as', as against 'either/or' in order to recognize a fundamental multi-dimensional reality that needs to be investigated using much higher degrees of specificity, location and overlap. In the period since 1989, there have been simultaneous shifts in different directions in global dispositions of economic political and cultural power. The early expectations that the collapse of the Soviet Union would usher in a singular hegemony of the USA and its guiding capitalist and democratic ideologies have been increasingly overturned during the contemporary historical period. In the twenty-first century, we have a renewed bipolar Cold War, in which American and Russian spheres of influence, particularly over Europe, have emerged. At the same time there has arisen a more complicated situation involving new 'great powers', most strikingly Japan, China and India, and other nation states of growing potentiality, including Brazil and Indonesia, Turkey and Iran and Nigeria and South Africa. The result is a multiplication of combinations in global power relations that produces new patterns of shifting influence and alliance among a much greater number of possible permutations that require much closer attention and investigation. One implication of this is the potential identification and comparison of 'global regions', based on a combination of historical/cultural elements and geopolitical power relations between dominant regional states that were once, in many cases, colonial territories, but, today, are themselves neo-imperial contestants in their own global localities. An outstanding instance is that of contemporary West Asia, where Euro-American and Russian interventions have been both inspired and complicated by the presence of proxy agents of Turkey, Iran and Saudi Arabia.

JR: How does this position help to clarify our understanding in general terms, but, more specifically, in relation to contested, divided and vulnerable cities?

MS: Contested, divided and vulnerable cities provide a highly illuminating instance of the gulf in interpretation and understanding that is associated with either/or confrontation in opposition to the as-well-as view of cosmopolitan thought and practice. Particularly in terms of cosmopolitical thinking, the most vulnerable – cities subject to the most intensive and expensive social and spatial disintegration involving armed violence from within and without – make evident the role and impact of a multiplicity of political manipulations of collective cultural identities in aggravating and intensifying extreme outcomes, up to and including what has come to be called 'urbicide'. To deal with this multiplicity requires going beyond classical binary understandings such as that very basic dialogue or dialectic between 'agency and structure', between material and symbolic realms. Systems of multiple interrelationships and causal connections can become overwhelmingly intricate, yet we need to recognize that issues of war and peace, normality and dislocation of life, society and space require an essential multi-dimensional analysis and explication. Needing to deal analytically with four or five such dimensions is rendered more difficult by attempting to present them in the linear form of a written exposition. Over a long period of time, I have come to recognize repeated and irreducible simultaneous presence in almost all cases of macro dimensions of economy and polity, social and cultural, and spatial clusters of variables whose reciprocal interactions provide crucial insights into the processes of reinforcing 'cumulative causation' that power the dynamic paths underlining the trajectories and outcomes that are of greatest interest in any field of study, policy, planning and intervention.

In terms of the contested, divided and vulnerable cities these dimensions are clearly constitutive: economically, the dynamics of neoliberal capitalist globalization ensuring the persistence of sustained 'horizontal inequality' between different communities, where extreme poverty exists side by side with extreme affluence on a tiny minority of the most favoured; politically, the dynamics of compromised sovereignty, government authoritarian or dictatorial regime involving administrative corruption and denial of public accountability; socially, the combination of discrimination, alienation, exclusion and persecution of disadvantaged and minority groups by a majoritarian dictatorship; culturally, the political mobilization and hijacking of collective cultural identity groups, involving racism and pernicious 'othering', xenophobic nationalisms, exclusivist ethnicities, fundamentalist religion and other rejectionist formations; and, spatially, the erection of armed barriers within the urban area, rigid segregation of groups by territorial division, expulsions, 'ethnic cleansing' and denial of participation in public space, accumulate one upon another to produce a spiral of punitive causation that can be incisively interrogated. On this basis, it can also be responded to by a countervailing nexus of cosmopolitan interventions supporting the advancement of civic consciousness, inter-group association, civil society cooperation, reciprocal recognition of diversity among cultures, increased resilience against violent attack from within and without and reaching out to transnational and global networks which transcend national, ethnic, religious and linguistic limitations.

JR: In what way does this edited collection of essays provide a positive contribution when seen from a perspective of cosmopolitan urbanism?

MS: This collection of essays and case studies has contributed further to a more comprehensive coverage of contested city situations that takes in the experiences of cities beyond the classical cases from the Global West and North. The 15 cities for which specific themes and local dynamics are covered adds to the growing academic and policy literature treated in previous collections such as Somme on 'At War with the City' (2004); Schneider and Susser, *Wounded Cities* (2003) and Bollens, *City and Soul in Divided Societies* (2012). Highly relevant also are Anthony King on *Re-presenting the City: Ethnicity, Capital, and Culture in the 21st Century* (1996), his *Spaces of Global Culture: Architecture, Urbanism, Identity* (2004), Edensaw and Jayne, *Urban Theory Beyond the West* (2012); Sanyal on *Comparative Planning Cultures* (2005) and Healy and Upton, *Crossing Borders: International Exchange and Planning Practices* (2010). The particular vision of urban geopolitics is a well-developed and well-timed contribution to what is now, I would argue, increasingly urgent and crucial enquiry in which the comparative perspective is greatly needed, given the dramatic and desolate perspective with which we are presented in contemporary world history, considering the total lack of effective response to disintegration and destruction in a lengthening list of cases from Sarajevo to Aleppo, where yet another instance of ongoing 'urbicide' is continuing as we speak.

Afterword

Lineages of urban geopolitics

James D. Sidaway

Where and when might an urban geopolitics begin and end? The chapters here have examined politics in colonial cities and a series of post-colonial conflicts. Ethno-territorial conflicts loom large, but so do the politics of development and confessionalism. In these multiple ways, urban geopolitics might be interpreted as a broad synonym for urban political geography (see Rossi and Vanlo, 2012), anticipated in prior work on cities and conflict (such as Boal, 1969), the geography of urban politics and the wider politics of urban planning (Kasperson, 1965; Mossberger *et al.*, 2012). As the preceding chapters testify, the range of these intersections is broad, defying easy generalization. They demand we historicize. The European overseas empires were tied to distinctive modes of the production of city spaces, commerce and security, so that, for example, bunds and cantonments became emblematic features of British imperial urbanism (Taylor, 2002), along with notions of race and colonial space. Likewise, East European and Soviet communism yielded relatively distinctive urban spaces (Smith, 2015): "determined in the complex interplay between political dictate, expert knowledge and bureaucratic norms and practices" (Kulić, 2016: 9). The consequences of American hegemony for the structuring of the urban have been equally profound. American-pioneered automobility constructs city space around highways and cars. Cities become fractured and connected by roads and economies and polities configured around petroleum (Campbell, 2005; Huber, 2013). Associated with these are what Don Mitchell (2005) called "the SUV model of citizenship". This was multiplied through American hegemony and is evident through large swathes of the Global South, reaching extreme forms when fuelled by petrodollars, as in Riyadh, Doha, Lagos, Luanda or Tripoli. Exceptions, such as Pyongyang, prove the rule.

However, references to urban geopolitics multiplied after September 11, 2001. The appearance of an edited collection on *Cities, War and Terrorism: Towards an Urban Geopolitics* four years later (Graham, 2004) and a subsequent monograph on *Cities Under Siege: The New Military Urbanism* (Graham, 2010) set the terms of the debate around urban geopolitics, illuminated by the 'war on terror'. According to *Cities Under Siege*:

> More and more, contemporary warfare takes place in supermarkets, tower blocks, subway tunnels, and industrial districts rather than open fields,

jungles or desserts. All this means that, arguably for the first time since the Middle Ages, the localized geographies of cities and the systems that weave them together are starting to dominate discussions surrounding war, geopolitics and security… the prosaic and everyday sites, calculations and spaces of the city are becoming the main 'battlespace'.

Ibid.: xv

However, there is a longer history of encounter between cities and the geopolitics through and beyond the twentieth century. This encounter is especially visible in sieges, such as Huê, Stalingrad, Khorramshahr or Aleppo, or blitzkriegs like Chongqing, Dresden, Hiroshima or Coventry and partitions such as Berlin and Nicosia. But the interface between geopolitics and the urban has diffused into everyday planning of metropolitan and post-colonial cities, as several chapters here testify; the late twentieth century saw numerous cities become battlespaces, most devastatingly Baghdad, Grozny, Kabul, Mogadishu and Sarajevo. Citing cities that have also cropped up in chapters here, Laurent Gayer (2014: 4–5) notes how:

The lasting contribution of 'political' and 'criminal' violence – two categories that cannot be taken for granted – the fabric of Karachi has earned the city the reputation of a 'South Asian Beirut', drifting towards chaos. As the memory of the Lebanese civil war faded away among audiences, this analogy dropped away in favour of a more global construction. Karachi earned the title of 'the world's most dangerous city'.

Dystopian visions – fortified with gated communities, compounds, fear, surveillance and barricades – are rightly centre-stage in many narrations of urban geopolitics. And while some of the chapters here have also focused on these phenomena, that many chart other imbrications of the urban and the geopolitical yields recognition of diverse encounters. They signal how epistemological and ontological status of the urban and the geopolitical (both of which signify many things replete with ideology and as categories of practice) present us with a complex analytical picture. Mao once said that the countryside would surround the cities – his strategy for China's revolution. And there have been cases when half or more of a city's population flee or are forced out of town, as in Phnom Penh in 1975 or Bangui since 2013. However, today it has been claimed that everywhere the countryside (along with the rest of the planet) is urbanized. So, in this notion, the urban becomes planetary (Brenner and Schmid, 2015). Is it a corollary of planetary urbanization that urban and geopolitics are homologous or equivalents? And, if so, what subaltern variants and alternatives emerge? *Urban Geopolitics: Rethinking planning in contested cities* has offered creative pathways into these expansive questions.

James D. Sidaway
Singapore, 20 December 2016

References

Boal, F. W. (1969) Territoriality on the Shankill–Falls divide, Belfast. *Irish Geography*, 6: 30–50.

Brenner, N., and Schmid, C. (2015) Towards a new epistemology of the urban? *City* 19(2–3): 151–82.

Campbell, D. (2005) The biopolitics of security: oil, empire, and the sports utility vehicle. *American Quarterly*, 57(3): 943–72.

Gayer, L. (2014) *Karachi: Ordered Disorder and the Struggle for the City*. London: Harper Collins.

Graham, S. (2004) *Cities, War and Terrorism: Towards an Urban Geopolitics*. Oxford: Wiley-Blackwell.

Graham, S. (2010) *Cities Under Siege: The New Military Urbanism*. London and New York: Verso.

Huber, M. (2013) *Lifeblood: Oil, Freedom, and the Forces of Capital*. Minneapolis: University of Minnesota Press.

Kasperson, R. E. (1965) Toward a geography of urban politics: Chicago, a case study. *Economic Geography*, 41(2): 95–107.

Kulić, V. (2016) The builders of socialism: Eastern Europe's cities in recent historiography. *Contemporary European History*, November: 1–16.

Mitchell, D. (2005) The SUV model of citizenship: floating bubbles, buffer zones, and the rise of the 'purely atomic' individual. *Political Geography*, 24(1): 77–100.

Mossberger, K., Clarke, S., and John, P. (2012) *The Oxford Handbook of Urban Politics*. Oxford: Oxford University Press.

Rossi, U., and Vanlo, A. (2012) *Urban Political Geographies: A Global Perspective*. London: Sage.

Smith, M. B. (2015) Faded red paradise: welfare and the Soviet city after 1953. *Contemporary European History*, 24(4): 597–615.

Taylor, J. E. (2002) The bund: littoral space of empire in the treaty ports of East Asia. *Social History*, 27(2): 125–42.

Index